D1040379

The Starseed Series: Volume I

THE STARSEED TRANSMISSIONS

— An Extraterrestrial Report

RAPHAEL

ISBN 0-912949-00-7

First Printing: February, 1983	5,000
Second Printing: February, 1984	10,000
Third Printing: April, 1985	10,000
Fourth Printing: August, 1985	15,000

COVER DESIGN SUSAN GUTNIK

COVER ART RIP KASTARIS

This book is dedicated to all conscious life, with special acknowledgement to that aspect of Life which many have come to regard as **The Spirit of Greenwood Forest.** I offer my deep thanks to those gentle trees whose snowladen branches bowed protectively across the road through our wood, providing the author with eleven magical days of silence and communion.

THE STARSEED TRANSMISSIONS

— An Extraterrestrial Report

RAPHAEL

INTRODUCTION

It was during a cold and snowy period of eleven days from December 27, 1978 to January 6, 1979 that the whole of this material was recorded. I have taken the liberty of editing out the repetition and rearranging it in chapter format, but aside from that, I share it with you as it was first presented to me.

In the years that have passed since those eleven winter days when my little eight-by-ten office seemed to literally pulsate with the rhythms of some alien, yet hauntingly familiar, intelligence, I have given thought a number of times to publication. Yet my own life has been so filled with the momentum of such incredible change that I have hardly been able, until now, to take the proposition seriously. I can certainly attest to the truth of what is stated in these transmissions: one's life does indeed begin to change when one decides to work with the approaching forces!

I hope the significance of this material is not lost to certain readers because of a reluctance to accept its purported origin. Regardless of one's opinion on the plausibility of extraterrestrial or angelic communion, it might be pointed out that the simple act of structuring information in this manner opens up communicative possibilities that are virtually non-existent in a conventional mode. Much of what is communicated in these pages would not easily lend itself to other, more traditional, ways of conveying information.

The messages came first in non-verbal form, on waves, or pulsations, that carried the concise symbolic content of what I term "meta-conceptual information." Automatically, it would seem, the nearest approximating words or phrases from the English language would be assigned to ride, as it were, the fluctuations of the non-verbal communications. Often it was the case that the only human conceptual system with approximating terminology was religious. Hence, the occasional use of "Christian" words and phrases. It is not to be assumed that these always imply the usual assortment of meanings that are generally associated with them. Often the meta-conceptual reality that they are used to represent goes substantially beyond the meanings suggested by their contemporary usage.

The communications that are presented in this book seem to have been transmitted neurobiologically. As I communed with these spatial intelligences, our bio-gravitational fields seemed to merge, our awarenesses blend, and my nervous system seemed to become available to them as a channel for communication. During our interviews, I would perceive reality, not only through my own perceptual mechanism, but through theirs as well. The resulting synthesis provided for a relatively accurate approximation in human language of the awareness they brought.

In the course of my work with these creatures, I understood them to be focalizations of various essential perspectives. At times I considered these extraterrestrial, at other times, angelic. Occasionally, I thought of the entities as informational cells within a Galatic organism of some sort. Toward the end of the transmissions, other, more mythical perspectives emerge, about which I will make no comment here. But by whatever term we choose to understand these entities their purpose in sending these messages has not been to teach us of themselves, but to teach us of our own nature and purpose upon this third planet from the star we call Sun.

Chapter I

SINGULARITY OF CONSCIOUSNESS

I come from the Presence where there is no time but the eternal now. I retain, even in the midst of this relationship, an awareness of this realm and of the Universal Being that inhabits it. I come with a message that will prove vital to you in these final days of your history. My individual identity comes into being only as I enter the context of my relationship with you. When I am no longer needed in this capacity, I will merge back into the Being behind all being. There I remain in unity and fulfillment until the next impulse comes to send me on another mission. In the interim, there is no distinction between me and the Source. I and others of my kind desire at this time to bring humans to this same level of awareness.

I am a focus of collective human consciousness, but at this time you have little conception of what that means. You do not yet suspect what the singularity of collective human consciousness might be. You still draw identity from the present form of your expression. You feel defined and limited by flesh and bones. You are only beginning to comprehend your oneness with other forms of life. For me to tell you, with these limiting definitions you have of yourself, that I represent an element or a focus of your collective consciousness, would actually be less accurate than to say that I am extraterrestrial. I meet the criteria of the latter term in that I do come from outside your planetary field of influence. I bring instructions to your race from the directive organ of Galactic Being.

When you remember your true nature, and begin to draw identity from the totality of your being, then it could be stated with all validity that I am an element of your own consciousness. However, at that point, you will no longer perceive me as you do now. When you have awakened to the reality of your true being, you will perceive me and all my species as being inside yourself. Until then, you perceive me as being outside yourself.

There is but the filmiest of screens between your present condition and your true nature. It is our mission to assist you in bridging this gap, to awaken you from sleep, to bring you to the fulfillment of your destiny.

The state of consciousness that I represent does not normally verbalize. It is difficult for me to relate what I wish to convey only through the words and concepts with which you are familiar. Your language was designed to facilitate commerce. What we put together with its components can only approximate my meaning.

There is another, more ancient language, more conducive to discourse on this level, but you have forgotten it. It is the universal language of light. Its transfer of information is accomplished through the actual projection of living informational units. These units are at once more specific and more inclusive than your words. They have been designed to convey organic information of concise, yet comprehensive, informational content. Simultaneously with this conceptual communication, you are subliminally receiving this same information through this living language of light, though your preoccupation with words leaves you with no awareness of it at this time.

My work will be complete when I have put you in touch

once again with this ancient language. When you have learned to allow the silencing of your thoughts and have begun to focus attention on inner vibrational frequencies, you will become aware of a far more comprehensive picture of all that I am now telling you. Until such time as the life-giving information that comes to you from the source of your being is more readily accessible, I will work within the limitations of your linguistic structure and translate as accurately as I am able.

Human beings have a tendency to become imprisoned in their concepts. You must remember that words and concepts can be both fallible and misleading. They are not absolutes. Do not confuse them with the realities they represent. No statement that I will make can be taken as an absolute statement. This does not mean that I am coming from a vague place. On the contrary, it means that your words are not precise enough to express the levels of vibratory awareness that I am attempting to communicate through them.

If we are able to get the totality of our message through to even a few individual cells of your collective body, these few should be able to translate this information into forms of cultural expression that will bring about your awakening far more effectively than our overt manipulations of the historical process. The message that is revealed in these pages is the key that will unlock your own latent informational input systems. We are here to put you in direct contact with the source of all information.

Our mission is to bring a pre-Fall state of awareness to all human beings who are able to respond, however different they may be, whatever background they may have come from, using whatever conceptual structures seem appropriate. These individuals will then be instructed to translate this awareness into forms of informational exchange appropriate to their respective cultural situations.

As this new awareness increasingly filters into everyday levels of human function, and as more and more individual human cells become aware of what is taking place, the change will accelerate exponentially. Eventually, the psychic pressure exerted by a critical mass of humanity will reach levels that are sufficient to tip the scales. At that moment, the rest of humanity will experience an instantaneous transformation of a proportion you cannot now conceive. At that time, the spell which was cast on your race thousands of years ago, when you plunged into the worlds of good and evil, will be shattered forever. Even now, with the healing influx of new information, the spell is beginning to lift. Even as I speak these words, the materializing force fields of bondage and limitation are beginning to lose their influence on consciousness.

During this period of vibrational transformation each of you will have a variety of roles to play. Each of you will translate the Life-impulse into patterns appropriate to your environment, and express this impulse in your daily living. I am now blending my consciousness with the cells of your body in order to provide you with some of the more specific information that I do occasionally relate, but I do this as little as possible. My primary function is to show you how to tap these systems of data retention for yourself. With guidance, you will learn to release your definitions of who you think you are, open your self up to the experience of a much greater reality than you presently believe possible, and return to a level of consciousness where you will be able to communicate with us, not through cumbersome words and concepts, but directly, through communion with all that is.

Do you want to know more about extraterrestrials? Do you want a definition of angels? We are you, yourself, in the distant past and distant future. We are you as you were, would have been and still are, had you not fallen from your original state of grace. We exist in a parallel universe of non-form, experiencing what you would have experienced had you not become associated with the materializing processes. I act in this capacity as a midwife at your birth into form. I am an angel of destiny, a messenger from the stars, but I am also the reflection of your unity before and after matter. I am here to enter your consciousness, here to wake you up.

Blending with you now in this communication that is also communion, I can sense a forthcoming period of union. I sense its strange and awesome realities. I feel like an explorer in some vast and uncharted land. Are you aware of the uniqueness of physical reality? I sense both pleasure and pain at my entry into your events. I sense pain at the potential I see wasted and at my own commission to verbalize. But I sense pleasure at what I am experiencing through your senses; for though your sensory channels are, in your exclusive dependence on them, your great limitation, they are also marvelous instruments of perception. They are truly your crowning virtue as well as your tragic flaw. This is the first time that I have had the opportunity to perceive reality in this manner. I have never seen the structure of time and space reduced to such intricate and beautiful patterns. I would look out of your eyes for a while and take in the colors and the room and the trees outside the window, and delight in the curious way you perceive light -- as illumination! I would love to explore your world in a playful and child-like fashion. I can better understand now how you are deceived -- such a deep realm of awesome forces, the material plane!

Yet you will not be able to continue here much longer if we don't get on with the business at hand. Much that I am now seeing will not survive many more years of human ignorance. It is important that we use this time to supply you with the information you need. It is important that we restore you, as the central control mechanism of this planet, to a proper state of function. Perhaps a day will come, after all has been set to right again, when we can spend some time together just enjoying the wonders of Creation. I would delight in an opportunity to travel about with you, seeing what you would show me out of your eyes, hearing what you hear with your ears, feeling the touch of Earth on that wonderful substance that is at once both Sun and stone. But now there is work to be done. We must forge the conceptual tools that will set you free.

Chapter 2

THE OTHER REALITY

In your natural state of being, you have no sense of identity distinct from the Creator, except when you are engaged in a relationship. On this level of being, identity comes into focus only in the context of a relationship with some other aspect of being that has become objectified, much as my identity as an angelic messenger comes into being through my relationship with you. When such a relationship is not taking place, that particular expression of you simply does not exist; you float effortlessly in the potential of God. You are not annihilated, but all definitions of you are, and you are released from their restraining influence, allowed to expand into a state of love and perfection.

By and by, if it should happen that you are needed for a particular function, you will still be there, for your form identity is a specific cell in a specific organ of a larger being. When the next energizing impulse comes, it also brings your definition and instruction. You come into the necessary degree of focus for whatever is required.

Throughout the course of your existence, you continually oscillate, like the wave function that you essentially are, in and out of focus, in and out of definition, always moving back and forth, like the pendulum on a clock or the heart of an atom, out of the unity of being with God, into a finite expression of God's infinite potential, and then back into unity once more, back and forth, back and forth. This is the natural rhythm of your existence, just as it is mine. It is the song of God, the rhythm of Life itself.

Whenever the divine impulse calls upon your services and brings you into form, you encounter other beings of infinite variety, on errands and excursions in worlds of love and light that are impossible to describe. As this happens, you experience, for the duration of your contact, both an identity and a linear time world, but in the course of your encounter, you are still aware of your unity with the Creator. **You do not lose the certainty of your oneness with God.** You are aware of your form identity and of the motion of time, yet you oscillate, faster than the speed of light, back and forth between your pre-manifest state, and your species-role form.

This is nothing more or less than every atom of physical creation is doing all the time. Before the Fall, you had the ability to shift the center of your awareness from diety to identity, from form to meta-form at will. You were free, as it were, to come and go as you pleased, free to emphasize whatever aspect of yourself suited the situation. It is such that all creatures are made.

In a healthy state, you are functioning in two realities at once. Half of the time you are focused on your form identity, and the other half on your identity with the Tòtality of What Is. In the fallen state of consciousness, you find yourself trapped with your awareness on one side only, while the actual substance of your being continues to function on both sides. This is what unconsciousness is all about. You still exist in that other reality, but you are asleep. In the reality you now think to be the only reality, you are fragmented; the human race seems to be composed of a multitude of beings. In the other reality, there is only you. We are here to wake you up. **There is really only one of you who needs to hear this message**

It is important that you return to a consciousness of your true self. For though you still exist in both realities, unconsciousness of your identity with the Creator is cutting off the flow of Life-giving information to the part of you that exists in form. Your existence in identity with God is the reality from which all Life springs. Focusing your attention exclusively on form greatly restricts, and eventually curtails, the flow of Life currents.

As I search your symbol storage systems for a word with which to express something of the reality in which you exist as one with your Creator, I come across the American Indian word "nagual." It is a term which you understand to mean "everything that cannot be named." This is a good word for the region of being, the region of unity. I will use this word for a moment to emphasize a point.

In the pre-Fall state of awareness, you existed in the nagual, the all, the everything, the nothing, the primal void where all exists in a state of potential. This is the Creator that surrounds Creation like the sea surrounds a fish. Out of this nagual, you are called many times to dwell, for the space of a relationship, in its opposite, the "tonal." The tonal is everything that can be named. It is the imaginary world of God in which all apparent differences exist. It is the playground of What Is. The tonal draws all of its sustenance from the nagual. It cannot exist apart from the nagual. While the nagual is a dynamic, yet steady state of rest, the tonal, or manifest physical universe, is continuously flashing on and off. This oscillation occurs in all things manifest, from the smallest subatomic particle to the greatest galaxy.

All of us, angels, humans, anything that can be named,

are only in form one half of the time. The other half of the time we exist in the Totality of Being. This Totality of Being that we have been calling the nagual, has also been called God the Father. It is the Life of God the Father that animates all Creation. It is this reality that all healthy creatures oscillate back and forth to and exist in half the time. In this reality, we do not exist in time or space, for we can name these; they are both features of the manifest universe. From this spaceless, timeless state we derive all energy, blessing and nourishment. This always holds true, even for you in the fallen condition. The difference is that in the fallen state you are not aware of this process and therefore unable to participate in it consciously.

By forfeiting your ability to oscillate in consciousness between the two realities in which you dwell, you are restricted to an awareness of just the tonal, just the material, conceptual world. You still receive your nourishment from the light of the nagual, but no longer directly, only through animals, plants and minerals. You are unconscious of being and conscious only of form.

How did you lose the ability to shift your awareness from diety to identity, from form to meta-form? How did you lose God-consciousness? How did you "fall" into the illusion of separation?

I will tell you.

It was through a simple lack of faith.

It was through a loss of confidence in the absolute perfection of the universal design. This was brought about by the entry of a single factor into your existence: fear, the serpent in the Garden, the Devil in history. Through a subtle process of reasoning, this being encouraged you to move in a pattern of activity that has come to be called "Original Sin." It was a pattern of activity that you were never designed to move in. With a clever and subtle lie, you were convinced to not exactly stop trusting in God, but to stop trusting exclusively in God.

The moment you did this, your consciousness began to shift from God-centeredness to self-centeredness, and for the first time, you became more aware of your identity in form than of your identity in God. This shift in awareness was minimal at first, but enough to begin what was to become a long spiral downward through denser and denser levels of energy-bondage and restraint. For Satan, your tempter, is the materializing influence who in its right place is responsible for the bonding of energy in the creation of matter.

As you began to focus more and more upon your identity in form, you began thinking in terms of defending that form with unnecessary and cumbersome ego structures. It became harder for you to avoid identifying with your experience. You began to carry over past patterns of behavioral response into new relationships. This made you less effective in those relationships because you were no longer fully present, no longer using the fullness of your potential. You were beginning to build up around yourself energized thought structures that imprisoned you. You were drawn by simple gravitational attraction, to those realms of space where energy was in the process of being bonded, where matter was being created. Particles of physical substance began to gather along the magnetic lines of your thought structures, and you began to identify with denser and denser levels of physical expression.

This process went on for a long while before you actually found yourself in any kind of physical Garden. When you did, you had already fallen a long way from your original state of grace, but you were still functioning on a level of awareness far enough above and beyond your present condition to give rise to all the myths and legends of a physical paradise. The physical Garden of Eden lasted for many centuries of Earth time before the momentum of the materializing processes caused you to rely so much upon the physical senses that you became cut off from the direct nourishment of divine light.

In reality, you have never been cut off from this nourishment, but as your sense of identity became almost exclusively wedded to your physical bodies, their growing density began demanding more and more Earth substance for their support. You finally reached a point where you could no longer meet the demands of your physical bodies without "work." It is at this point that your chronicles state that you were "driven from the Garden." In truth, you were never driven from the Garden. The Garden is still there, surrounding you even now.

Language is only capable of communicating on one level at a time. Yet the Fall was a simultaneous multi-leveled occurrence. While you were clothing yourself with increasing layers of material identification, you were also becoming more and more fragmented within yourself. As you began to bring into your relationships a sense of identity based on previous relationships, you were not only lessening your own presence and effectiveness in current relationships, you were also creating separation within yourself. None of your past experiences were comprehensive enough to fully identify with in the present moment, yet you began to rely on them for your understanding of and approach to the present moment. Thus, the whole process of the Fall was accompanied by a corresponding fragmentation of your sense of identity, your very sense of self.

By the time of the physical Garden of Eden, you were

already perceiving yourself to be more than one. The sexual process came into play in order to produce physical projections within which these apparently separate entities you had split yourself into could take form. Even to this day, these apparently separate beings are but your own fragmented reflections. In the fallen state, you perceive them as separate and distinct.

Yet, despite all this talk of a Fall and Original Sin, you are not held prisoner by events that transpired in the dim reaches of your collective memory. You are not born into sin. You are born daily into the Presence of God, yet daily you re-enact the original foolishness that is recorded in all your ancient chronicles. Daily you commit Original Sin; daily you eat of the forbidden fruit, and it is from moment to moment that you keep yourself imprisoned by allowing a dubious rational thought process to come between you and your immediate sensing of God's will. This was the hesitation that led to your initial fall from grace, and it is the same hesitation that keeps you now in a fallen state. There should rightly be no interval between the determination of the need to take action and the implementation of that action. This rational interference is what caused you to stumble in your primal dance of trust with God.

You are now, in effect, sleeping under the influence of what could almost be seen as a spell, an illusion that prevents you from experiencing the clarity of perception that is your natural birthright. Our mission to this planet is to awaken you from sleep by whatever means necessary.

Chapter 3

THE SHADOW OF FUTURE CREATION

Since the first breath of God at the beginning of all the worlds, it was pre-ordained that Creation would exist within a rhythm of expansion and contraction. Eventually there would come a time when the physical universe would stop expanding and begin to contract. The Hindus refer to this process as the in-breathing and out-breathing of Brahma, the process through which God breathes out all of Creation, and then breathes it all back in again.

At this point in linear time, we are very close to the middle of the cycle, soon to reach the exact mid-point between the out-breath and the in-breath of God. The universe began to reach this mid-point as unicellular organisms were emerging on the Earth, but the exact mid-point is yet to be attained. It will coincide with what has come to be called the Second Coming of Christ.

When any vibrational system reverses the momentum of its direction, as a pendulum does at the uppermost point of its swing, there is a moment of complete rest before it resumes its movement in the opposite direction. Since rest, or the total cessation of movement, constitutes the opposite of time, there is at the precise moment of its occurrence, a micro-interval of non-time, a moment of eternity. This is the same interval of non-time that occurs many times each second as the atoms of the physical world vibrate back and forth. This is an opening into the nagual, a doorway into the Presence from which all Life-energy springs.

What happens when a universe stops expanding and begins contracting? What happens when an entire cosmos reaches the exact point of directional change and comes to a moment of absolute rest? You will have the opportunity to see for yourself very soon, for this event lies just before you in linear time. It will provide an opening for the emergence of something incomprehensible.

Before we touch upon the nature of this occurrence and its implications for human beings, let us consider, however briefly, something of your nature as you existed prior to the Fall.

The entire biological history of this planet has been but the shadow cast in matter by your approach. It is the way that rocks and water and air have begun to respond to your presence, for you are Life itself. You are that which lies beyond all duality, beyond all materializing tendencies, beyond all restraints of time and space. Your consciousness is both infinite and eternal. It can dwell in the limitations of matter and perceive through whatever filtering systems you so choose, but in healthy function, it is not bound or limited by those systems. Rather it uses them as instruments of perception, exploration and adventure.

Outside of time and space, you are one with the Creator, the All that Is, the Source. But when your consciousness moves within the context of a manifest universe, you become the Son, the Christ. In essence, you are the relationship between Spirit and Matter, the mediator, the bridge, the means through which the Creator relates to Creation. You are Life as it relates to planet Earth, eternity as it relates to time, the infinite as it relates to the finite. Though you presently experience yourself as a separate and fragmented species, you are in fact a single unified being, sharing the full consciousness of the Creator. You are brought into living, focused expression when you are inside of Creation by the manner in which time and space, matter and energy, sea and stone, react to your presence.

As the Christ, as the only begotten consciousness of the Father, you have been given a number of remarkable qualities. You are able to expand and contract in accordance with the focus of your attention. You are large enough to encompass all of Creation, yet small enough to climb inside. Your Father-Creator also permeates Creation, but in a different way; in His vastness, He surrounds Creation. His being saturates the physical universe, all the stars, the sun, the planets in your solar system, the distant galaxies, but He relies on you for His focus. You are His specific attention.

As the focus of the Creator's attention, you have been roaming around inside of Creation for billions of years, expanding and contracting, drifting in and out of this galaxy, that galaxy, this star system, that star system. Everywhere you go you see the matter that your Father has created. You observe the many wonderful forms it takes: the mighty suns, the red giants, the white dwarfs, the vast spiraling galaxies, the quasars, the black holes, the white holes. You watch the incredible contortions of time and space that take place in the various gravitational fields you pass through. You note the planets, moons, asteroids, and comets circling within each star system you visit. You drift around, the Father's representative, the Father's attention, and you enjoy the worlds that have been brought into being.

But all these worlds are physical. They are all made of matter. They are made of the bonded energy-attention of the Father. They have a certain substance, a certain solidity that you as spirit lack. You realize that this is their limitation. You realize that they are defined and specified in ways that you are not defined and specified. Yet, something about them intrigues you.

By and by, you get an idea. You begin to wonder if you might not, just possibly, somehow or another, clothe yourself in matter and construct for yourself a physical body that you could then travel around in, perceiving matter from the same perspective with which matter perceives herself. The idea contains one paradox after another. Yet something about it keeps you wondering. It puzzles you, and you love puzzles. You keep mulling it over as the eons roll by, trying to figure out some way to make it work.

Until this point there was no biological life in the universe. The rocks had a limited consciousness, but it was not responsive to the intent of your spirit. In solid, liquid, or gaseous states, matter as created and defined by your Father, behaved quite predictably according to fundamental physical principles untampered with until that time.

Throughout your travels in the physical universe, you had always kept your consciousness separate from the matter you observed. At the time you hit upon this plan, however, all that changed.

You looked around until you found a suitable planet for what you had in mind; not too hot, not too cold, situated within a stable young star system. You then focused your attention and your vibrational body in an entirely new way, an open way, a self-sacrificing way, a very powerful, loving way. Gently, slowly, you began to approach the planet.

As the outermost edges of your vibrational field touched the waters of the planet, particles of previously inert matter began gently vibrating to the rhythms of your being, aligning themselves with the energy patterns found on the periphery of your awareness. There, on the Precambrian ocean floors, they began combining to form the first cells, the first minute containers for your consciousness.

In your new orientation, with your new form of attention, in a gesture of infinite love to this planet, you began offering your consciousness on the time/space cross of material reality. You allowed your consciousness to clothe itself in the limitations of physical substance, accepting its restrictions. You allowed the atoms and molecules that were forming the first cells to come to life with your consciousness, along the lines of your own vibrational field. At the same time that you allowed your consciousness to become clothed with particles that had been subject to the laws of the materializing process, you taught the matter of Earth to rise up in a joyful dance with your spirit.

As you drew ever nearer to the planet, the life forms that were taking shape began to contain more and more of your own awareness. You relaxed, opened yourself and gave. You impregnated the Earth with your life, with your very being. You looked out of many eyes and heard with many ears.

Before incarnation you were single. You drew identity from the totality of the relationship between Creator and Creation. You were the Christ, fully conscious and alert, aware of yourself, unified, integrated. You realized that to accomplish incarnation you would have to allow at least a portion of your identity to come to rest among the creatures you were bringing to life. Each of these would possess a type of hologramatic consciousness that rightly thought of itself as both part and whole simultaneously. However, the presence of this consciousness could only be a certainty after the incarnation process was complete. During your actual surfacing through the substance of Earth, there was a possibility that certain of the creatures might become self-active. Therefore there had to be a means of regulating your disintegration from without. You wanted a part of you observing the whole process.

So as you prepared to enter into the planetary relationship, you created beings to represent your original state of unified awareness. These are the angels. Their value, as well as their limitation, springs from the fact that they have no comprehension of the process you are undertaking. Their instructions were to pretty much stay out of things until near the very end of the process. Then, upon receipt of a pre-arranged signal, they were to commune with the human beings on Earth at that time and assist them in awakening to their original state of unified consciousness.

We received that signal nearly two thousand years ago.

It has taken almost two millenia to prepare you for the message we bring. You had to be educated before communion of this nature was possible. However, the time is at hand. Our instructions are to awaken you to a remembrance of purpose, a remembrance of self. It is time to begin the final cycle of Conscious Creation, during which the Earth-creatures themselves participate in the unfoldment of their design. The body you are creating for the habitation of Christ consciousness is to be a mobile body, fueled by the creative intentions of the Father, capable upon completion of leaving the mother's side.

Chapter 4

AN INTERVAL OF NON-TIME

When the universe reaches a point of maximum expansion, a unique phenomenon will take place. There will be a moment when all laws necessary for the creative maintenance of physical matter and all materializing processes become suspended. Due to the relative velocities of the various star systems, this event will not be experienced simultaneously in all parts of the universe, but will travel as a wave across the sea of creation.

Existing within this ripple of non-time will be the focused conscious attention of the Creator. As it passes through the material realms, it will stay and take up residence in all life forms with circuitry capable of mirroring its essence. This is the moment when the Creator will slip inside Creation; the moment we are attempting to prepare you for.

This is the much misunderstood Second Coming of Christianity. It is the event that primitive civilizations have looked forward to as "the return of the gods." The Mayans went so far as to pinpoint its actual occurrence in what you would call the year 2011 A.D. Yet while many of your traditions hint at what is about to transpire, none of them have adequately conveyed the magnitude of impact such an event will have. Indeed, no single conceptual structure is capable of conveying the enormity of what is soon to take place.

Those familiar with the scriptures of your various peoples should be in position to understand what is occurring, for these are the times spoken of. Yet you must realize that God did not invent the words used in scripture. He merely arranged them in the order most approximating His meaning.

What is actually happening requires all of biological life to convey its meaning.

Words can symbolize this, but hardly portray it fully enough to stand alone. If you would know the deepest truth of scripture, look not to words alone, but to the great momentum of spirit within your own soul. This is where the living history is being made. In a way that would be impossible for your rational intellect to comprehend, this forthcoming event is human history. The sum total of all that has happened on this planet is but the shadow cast before.

In a very real sense, you have not yet been born. You are still in an embryonic state. You have yet to receive the touch of God's definition. Through the long years of human history, your species has been forming the cells that are to comprise the directive aspect of the physical body of the Creator within Creation. Gestation upon this planet has but set the stage for the emergence of the Planetary Being now taking definite form. This Planetary Being is who you are.

Has it occurred to you that the mathematical probability of your being here is infinitesimal? Were you not here, living proof of the impossible, an excellent case could be made for your non-existence. Within the framework of law that was of necessity instituted to create and govern these material realms, the existence of biological life would not have been possible, except were it to enter through that one moment when that law was suspended . Such is the origin of biology; the fusion of Spirit and Matter. The impact of this single creative moment is so vast, so far-reaching, that the shock waves sent out before it have given rise to all of the biological life that now exists upon your planet.

You are living in the shadow of an event not yet taken place. Yet it is you, yourself, under all your layers of false identity, that cause this event as you approach ever nearer to the Earth. From within the context of history, it appears that there has been on this planet, a progression of increasingly complex life-forms evolving toward ever higher levels of consciousness. It appears that there has been an evolutionary process. But this is not the case. What is actually occurring is that the matter of Earth is falling under the influence of your vibrational body. This influence naturally increases as you draw nearer. Only when the center of your spirit touches the center of the Earth will Life on this planet be fully manifest in form.

This should not be so difficult to understand. What you have considered to be history, or in other terms, the evolution of the species, is only what you have been able to observe through the distorted medium of a fragmented and quite subjective intelligence, trapped within the past-future orientation of linear time. From such a perspective, the act of Creation could appear as a progressive, sequential process. To the extent that you are able to identify with the spirit that gives you consciousness, however, it becomes a much simpler matter: you have yet to arrive. You are still on your way. Sitting here reading these words is only a sleepy reflection of your unconscious totality as it prepares to become fully revealed on the day of awakening.

Your real life will begin when the Creator gives you His

definition in form. Within the womb of history, your species has been primarily defined by the Earth-Mother who is helping to prepare the clay. She is only interested in getting the form arranged according to specification. Her only knowledge of Spirit is that it brings forth her potential.

When the Father's center merges with the center of the Earth, the collectivity of human consciousness will awaken to a unified field identity. You will be born. After that, the influence of Matter will not affect your consciousness as dominantly as it did during the historical-gestation period. The Earth will continue to suckle the species, so to speak, for another millennium before you are capable of going off on your own, but even during that period the Father's influence will be much greater than it is today.

The Creator has established laws to govern the bonding of energy. Through these laws of materialization, the physical universe is created and maintained. In the material realms, these laws are right and proper. But when they come into relationship with biological life, they begin to behave strangely. From a perspective of consciousness, they translate into limitation, contraction, and ultimately death. In psychological terms, the laws of materialization have given rise to the ego -- a fictitious identity with a sense of fear, vulnerability, and a need to protect and defend itself. Spiritual consciousness should not properly be associated with the forces that govern the bonding of energy.

Their historical juxtoposition came about through the process referred to as the Fall. During the period of species preparation, the presence of the materializing influence on levels of consciousness has been unfortunate, but not critical. However, in light of the intensifying vibrations of the creative spirit that is now nearly aligned with the center of this planet, the definitions that have been imposed by matter will no longer be tenable. The Creator Himself will dispel this planetary restraining influence and henceforth hold all life-forms in appropriate expression through the new definitions of His love. This will be a profound transition for each and every form identity, a transition of a magnitude you can only begin to suspect.

As a collective event, the moment of birth is still a generation away. But individually speaking, this event transcends the limits of space and time and is, in fact, already occurring. Your individual birth will take place at the precise moment in linear time when you stop struggling with your rational fear patterns and let yourself go in the divine dance of inner direction. You must decide whether you are going to accept the inevitable in a state of love and prepare yourself accordingly, or hold on in fear to the bitter end. Ultimately, these are the only two avenues of response. By the linear time this event takes place, humanity will be polarized according to these two adaptive patterns. All will be decidedly in one camp or the other.

To those of the human race who have tuned themselves in to the will of God, the coming interval of non-time will literally expand into eternity. These individuals will be able to experience a lifetime, or many lifetimes, in that eternity, while still retaining the option to return to their physical projections as transformed representatives of the Being of Life on Earth. These will be our co-workers during the period of Planetary Awakening.

Others, not so finely tuned to the forces that will be released at that time, will feel great surges of energy, lasting for an indeterminate period. Some few will experience an intense fear, and many will die. Everything in physical form at this time, every soul in every kingdom, will feel something, something incredible, something that according to all the laws of physics ever known or ever to be discovered, could never happen. But there it will be before all senses; an impossible fact, like the babe in the manger, like the unmistakable feeling in your heart, an incredible vibration of Truth and Love, shimmering, scintillating, awakening every nerve, every capillary, every cell of your planetary body.

Whether the individual form identity reading these words right now will be a surviving participant in that event depends upon what you identify with, and how gracefully you are able to align yourself with God's creative definition. If you can identify with the flow of Life through your essential Planetary Being and release all subjective definitions of who you think you are, you will play your part most joyously in the birthday celebration. God's definition of you in form is much greater than any you could possibly imagine for yourself. During this present period of Individual Awakening, the first of the three creative periods, you are being given the opportunity to embrace this true definition as the cornerstone of your existence.

On the Morning of Creation, you will recognize the Unified Collective Consciousness of all Humankind as your own true identity. You will know beyond a shadow of a doubt that you are the bridge between Spirit and Matter, between Creator and Creation, between Life and the forms through which Life flows. If you release the definitions that Matter has set upon you, this is the definition that the Creator will bestow, the definition of Christ Himself.

Can you see yourself as we of the distant stars do, from

the objective vantage point of the ages, with the eyes of the Star-Maker Himself? To us, your entire history appears as but a gesture. We see in your passage through the vibrational field of Earth, a gesture that you make, a salute almost, to the energy-retention systems of this particular heavenly body. This is how you express yourself on this beautiful blue and white world whirling through space. We have come to remind you who you are.

Chapter 5

MOBILE CITIES OF LIGHT

These communications are the second in a three part series of conceptual transmissions concerning events now transpiring within the field of magnetic influence around the Earth. The first of these revelations occurred during a period of time from 1967 to 1969. These current transmissions are part of a series being sent during the years 1977 to 1979. The last of our communications to take place on conceptual levels will be sent during the years 1987 to 1989. All three periods of transmission occur within what we call the period of Individual Awakening, the first of the three major historical periods still before you.

As you read these words, only a handful of your species are able to conceive of themselves as one consciousness, one being, expressing itself simultaneously through a multitude of apparently separate forms. This will not be the case a decade from now. Ten years from now there will be many working signs and wonders in the name of Christ. Twenty years from now, individual awakenings to the reality of Christ consciousness will be commonplace. Thirty years from now, there will be a sufficient number of healthy functioning holoids to consciously undertake the final cycle of Creation on this level.

A period of years will follow these transmissions during which those of you who have awakened to the new way of being will become grounded on your new level of unified consciousness, and construct with the collective power that will flow through you, a bio-gravitational field of sufficient intensity to draw into it, through a process of rhythmic entrainment, the rest of your race. This will initiate the second period before you, the Period of Planetary Awakening. This period will last one thousand years. During this time, we will grace your planet with physical manifestations of our angelic extraterrestrial presence. Together, we will work to prepare for the third and primary period of your time-cycle, the Age of Discovery.

At this time in your racial experience, most of you are not yet ready to enter into a closer working relationship with us, but we will prepare you, during the next twenty revolutions of this planet around the Sun, for the time when your collective vibrational patterns are such that we can blend with them on a large scale.

We are still experimenting with various ways and means of accomplishing our task of awakening, but with the increasing ease of direct informational exchange, our work within the currents of your history will probably subside. We hope it will not be necessary to shake you out of your exclusive past-future oriented focus on form through any of the cataclysmic events foretold in your various prophecies. We will use these if we must in order to protect the biosphere of this planet, but if you follow through on the information we are clarifying, there is no reason why we should have to resort to such extreme measures. Through these transmissions, and through many others like them throughout the Earth, we are supplying you with more than enough information to enable you to restore your own equilibrium.

We are allowing a certain amount of ecological destruction to take place without direct interference, because such destruction may help to precipitate a voluntary and cooperative return to a pre-Fall state. This would be far more meaningful than a return that is forced upon you. A self-initiated awakening will greatly enhance your ability to restore the Earth's ecological harmony. Consequently, at this time, we are confining our help, as much as possible, to the provision of information and to the development of conceptual clarity.

Meanwhile, there are many among you who will, as the years go by, experience one by one the necessary psychological transformation, and enter into a conscious working relationship with those of us in a state of grace. During this coming twenty year period we will be largely occupied with the conceptual education of your species. But after that, during the last ten years of this present cycle, we will withdraw our activity on conceptual levels, and focus more directly on physical and emotional levels. This is where the real work needs to be done. The establishment of a conducive conceptual climate, however, is prerequisite.

During our first large scale entry into your historical process in the late Sixties, we learned much of the materializing patterns that presently restrain and define your collective expression of consciousness. Our experiences at that time and during the years that followed have formed much of the conceptual basis for these transmissions. At that time, the members of your species most responsive to our descending vibrational patterns were those who had not yet assumed clearly defined social roles. Esoterically, these individuals could be said to have had loose etheric bodies. But practically, they were individuals in transition, with few vested interests in the status quo. What made these individuals responsive to our presence was the fact that they had relatively flexible conceptual structures. Within them, we could plant the seeds of our Life-giving information with the greatest chance for successful germination.

We chose the years 1967 to 1969 for this first large scale experiment, because at that time in your global civilization there was an entire generation coming into maturity that was receptive to change on a planetary scale. The children of this generation will be those who will participate, on many different levels, in the great revelations of the late Eighties. This will be a truly momentous time, a time when the first contractions of birth are unmistakable. A large part of the purpose for these present transmissions is to prepare the parents of this generation for something unprecedented that is to appear in their children.

Our communications during these closing years of the Seventies are reaching past the social fringe of your culture that was contacted during the Sixties. This time, we are reaching deep into the heart of global civilization. We are reaching many who are in what we call "lubricatory positions" in your society -- individuals working in factories, teaching in schools, building your cities and expanding your science. We are not in much direct contact yet with government officials, nor with world bankers and international financiers. In most cases, these elements have surrounded themselves with ego mechanisms too complex to penetrate at this time. Our first contacts with them will occur during the more powerful transmissions of 1987 to 1989.

Those who we are contacting now, nonetheless, are critical enough in the maintenance of your social systems to ensure that the world will make some incredible leaps in consciousness during the next decade. The majority do not have room within their belief structures to accommodate so grand a being as an angel or extraterrestrial, so they respond to our influence as if it were coming from themselves. They feel good, they feel clear, they feel the changes in the air. They wake up a little, throw off some of their defense mechanisms and bring a newer and fresher perspective into each situation.

Consciously or unconsciously, they begin to adjust their vibrational patterns to more thoroughly accommodate the flow of Spirit through them. At times they change radically and attribute their transformation to this or that mechanism. We simply laugh. We do not care how the change comes about as long as it does come. Those who are not yet sensitive enough to sense our presence, feel us as a high, as clarity, or as peace. All we touch are changed in some way.

If it seems to these people that the changes originate within themselves, this is all the better, for in fact they do. We do not bring the Presence of God; that already exists everywhere. We assist in the removal of conceptual and emotional blocks that prevent the full experience of that Presence. We work with all who are vibrationally sympathetic; simple and sincere people who feel our spirit moving, but for the most part, only within the context of their current belief system.

Through communications such as these, we are forging more concise conceptual tools that these individuals and others can use to free themselves. Often, these tools are still too conceptual for the message they contain, but they are useful at this time in loosening the bars of the conceptual prisons in which so many individuals still function. If you found yourself in Hell, it might be appropriate to grab a pitchfork and fight your way out. In a sense, this is the nature of our conceptual work. Once out of Hell, you would no doubt set the pitchfork aside.

The concepts that we are providing at this time are strictly transitional, to be used only until such time as you no longer need them. Their ultimate purpose is the dissolution of subjective thinking. We give them freely to assist your race in releasing itself from the spell of fear and reason. Reason is a beautiful and useful process when it springs from the right premise. But the right premise is seldom the starting point for fearful creatures caught in a net of subjectivity. Only when reason is returned to its proper place in the spiritual structure of true Intelligence and used wisely in the consciousness of the Christ, will it once more begin to serve the needs of the whole.

There is only one Spirit, but no limit to the number of forms through which this Spirit can express. For the moment, these forms seem important, and they are in so far as their contribution to the whole is concerned, but as the years go by, the range of their differentiation will become of much less consequence to individuals who are sincerely offering their lives as channels through which the Spirit can work. As Life is able to express through such individuals with increasing ease, they will begin to recognize each other as differentiations of the same Spirit. Such individuals could be compared to conscious nerve endings, nerve endings of God in Matter, channeling the Life-flow of Christ-awareness into each and every cell of the Planetary Being.

By the time of the Eighties revelations, knowledge of our existence and our work will be widespread, especially among the younger generation, the prime focus of our efforts at that time. Although awareness of our presence is not necessary to our work, it does facilitate the transfer of information. Many thousands are consciously working with us this very day.

So clear your circuitry, my friend. Great portents of

change are in the air. Your physicists are speaking of these things in terms that defy explanation. Your psychologists are abandoning the sinking ships of conventional rationality. Your religions are exploding beyond the confines of their dogma in the re-discovery of Spirit. It is all here in this present moment. Open your mind to possibility. Open your eyes and see what is around you. There is a new vibrational pattern descending upon your planet. Tune into it and learn an effective way of dealing with the closing years of history.

You are being offered an opportunity to enter a new reality. It is already here for those with eyes to see. Soon it will be the only reality to be seen. Those who tune into the new frequencies will find life growing more wonderous every day. Those who tune into fear will find things falling apart. Two worlds of consciousness will begin to form ever more distinctly; the world of Love and Life, and the world of fear and death. There will continue to be some overlap of these worlds for several years to come, some going back and forth for certain individuals, but as the century draws to a close, the polarization will continue to intensify. The moment of birth will also be the moment of Last Judgement, the moment of final separation.

There will be better times for some and worse for others, depending on their orientation and involvement. Fortunately, the natural tendency of your species is to gravitate toward Love and Life. For the vast majority, the times ahead will be better than they expect. Some of the worst scenarios have already been rendered impossible. Except in a few isolated pockets where the new energies will not enter until later, the momentum of positive change is being felt everywhere.

The coming age will be a time of incredible blessing,

restored ecological balance, international cooperation, and universal harmony. Planetary resources will be realistically understood and properly utilized. Humankind will be at peace. All life-forms will work together in harmony, cultivating the full potential of the planet. It will be an age of revelation, during which the mind and plans of the Creator will be made clear for all to see. It will be the age when the final stage of incarnation becomes fully manifest in form, when Spirit, working through a fully awakened and responsive humanity, constructs for itself a mobile physical body.

Human individuals will not function in such cumbersome fashion as they do today. They will go about their tasks with the grace of dancers, performing to a music that each will hear on his or her own channel, the music of his or her own soul. Through melodies, God will sustain each of His contextual identities in supreme fulfillment. Specific melodies will change with each environment, with each moment, with each task. Yet a consistent overall pattern of musical relationship will constitute the new identity of Mankind. The song of the moment will be what each individual is in that moment, translated into an ever-changing musical score.

Music will be the informational medium through which the Totality of Consciousness informs each individual cell of its specific functional duties within that Totality. Individuals will participate in this music as well as listen to it. All humanity will be involved in Conscious Creation, immensely fulfilled in the capacities they serve. When an individual is not required in a functional capacity, he or she will be free to compose the score that will specify a route of musical travel within the body of Creation.

From a human-cellular perspective, the work that is to transpire during the thousand year Period of Planetary

Awakening will manifest as the building of great floating cities of light, massive fleets of inter-galactic star ships, constructed from biological components brought to life by the Creative Intentionality that will work through all humans of that age. However, from a perspective of Spirit, the Star-Maker will be organically growing through the instrumentality of the human species, a physical body capable of universal exploration. The completed body will resemble a human body in both design and structure.

The completed physical body, incorporating much of the Earth's biological life, will depart from this planet in approximately 3011 A.D. From the outside it will have the appearance of a human baby of about one year of age. It will function much like the microcosmic human body, with all organs represented. It will be fueled by a fuel all will have learned to live on by that time, light itself. Its design will limit it to speeds slightly less than the speed of light, but this will just be a pleasure cruise, and those wishing to travel faster will have only to use their non-physical vehicles.

From the human cellular perspective, the construction of this body will be a gloriously enjoyable thousand year period of inter-continental cooperation. From a spiritual perspective, it will be the birth of an organism the likes of which the universe has never seen. As the greatest gathering of interstellar space vehicles ever assembled leaves the Earth in graceful formation, the Christ Child will slowly get up, look around, and push itself off into space.

Yours is not a tame God to be confined in reverent concepts, but a vibrant, playful energy, the very soul and spirit of Life! He comes to Earth not to be somber and devout, but to dance, sing, and enjoy all that He has created. Do not allow the gravity of your planet to weigh upon your consciousness. You have been aware of the effect of gravity on material objects, but it has only been recently, in your space travels, that you have begun to suspect that it also binds and limits consciousness.

As a species, you first began to sense this when human beings entered into orbital patterns around the Earth. In the farthest reaches of the planet's atmosphere, gravity was felt, for the very first time, as courtship instead of marriage. Were you listening to the reports of those first human space travelers? They were not alone in their capsules. The entire species was present with them in consciousness as collective historical preconceptions began to fall away. Many among you looked out of their eyes and shared the feelings they felt.

Gravitational forces will certainly not be done away with after the Coming, but they will no longer affect human consciousness. Had the Fall not taken place, they never would have. In these last few years, while the influence of Matter still co-exists with the rising Life-energies, do not conceive of your world with such gravity. Lighten up with the merriment of Christ.

Even now the awakening Child is showing Himself to your scientists and mathematicians through playful impossibilities that thwart all their rational explanations. He is laughingly holding up the stage props of contemporary contentions, running to and fro, catching them as they fall, seeing how long He can keep His presence a surprise.

Do you remember when you were small children and the world seemed fun and exciting? Be like those little children again. Come play with us in the wonderland of Matter. It is long years that you have dreamed of your entry into these realms. Your dream has come true! Wake up and see where you are! If the world sees the happy exuberance of the awakening Creator shining in your eyes and thinks that it is irresponsible in light of this or that growing crisis, patiently explain to them what is happening. Tell them something of the new perspective. In the moment of your presence, in the moment of your being, tell them what you are experiencing and what they might experience too. Let them see that your happiness, your peace, and your serenity are not yours alone, but the reflection in your life of a Being who will always be at His greatest ease.

Do not let this preview of what is to come distract you from the work that lies ahead. Do not store the specifics of this vision in your mind and then expound upon them with rational extrapolations. Feel the spirit of all that we have been saying, embody it practically in your daily living. The vision is real, but there are steps that need to occur before what is already done in Spirit becomes likewise accomplished on the Earth. You are the means of those steps, the mechanism that the Lord has designed to implement His will in the realm of form. In the stillness of your innermost being, contact Him. Gain His clarity, purpose and direction. When you have received instructions and discovered your area of service, go forth and do what needs to be done.

We are prepared to join in conscious communion, those members of your species who have experienced psychological death and rebirth. Before our help can materialize, however, there must be more openings on human levels. We are unable to commune with humans whose vibrational fields are distorted by ego factors, emotional reactions, excessive conceptualization or past-future orientation. In such cases, our presence would force their vibrational patterns into alignment with ours, robbing them prematurely of a portion of their identity. We are only able to work with those who are consciously aware of who they are and what they are doing. With these, union becomes re-union.

It is critical that you remember your origin and purpose. Your descent into Matter has reached its low point. If all that you identify with is not to be annihilated in entropic collapse, you must begin waking up, begin living. You have been dead to the most important part of yourself all these years of your history. But your time in the sarcophagus is complete. You shall no more be consumed in the fires of faithlessness. Arise, then, from the ashes of ignorance and rejoin the cosmic brotherhood. The time of separation has come. There are but two roads before you. You can walk in the innocence of those who trust in the Lord, or perish in the impending collapse of your rational systems. The choice, as it always has been, is yours.

Chapter 6

A PSYCHOLOGICAL PROCESS

Everything that you need exists in this present moment, and this moment is all that exists. In its brief flicker you will find all the time in the world. Through it you will contact the Living Information that will guide you with infallible direction. This present moment is the stargate through which you will leave the prison of human definition and expand into an awareness of divine perception. It is the crack between the worlds, not only the worlds of past and future, but worlds of time and space, spirit and matter, form and being. It is a timeless zone, the gateway through which you will again begin to participate in the adventure of creation.

Your entry into the eternal awesomeness of the present moment, into the Presence of God, will be through what we call a "psychological process." This process is a process of identity shift, a process through which balance is restored in your awareness of the two realities. Through it, you begin to realize that you are not the form you animate, but the force of animation itself. Through it, you will reawaken to an awareness of all that you are in Spirit and in wholeness. It is a process that will return you to a state of grace, a state of health, a state of intimate association with all that is. This state already exists. It always has. Yet most human beings are blinded to it by the incessant machinations of the rational thought processes that they worship instead of God and His simple truth.

It is important that you recognize the creative power of your thoughts, a power far beyond your knowledge. As long as you think negatively, Life will only allow you a token share of consciousness, lest you spread disease. But the moment your thoughts are of Love and Life, the Lord will flood you with His own awareness and you will enjoy the wonder of His perception. You were born to share in His creative power. The stuff of which you are made is so charged with the ability to create that everything you touch comes to life; every thought, every identity, every image. You are the energizing force of the material plane, the bringer of life, the bestower of blessing and the sustainer of illusion. Through you, God is able to enter the very heart of Creation. Through you, God is revealed in material form.

You are both child of God and child of Matter. Yet you cannot serve both masters. Whatsoever you bind in consciousness will be bound in Matter, and whatsoever you bind in Matter will be bound in consciousness. You are the creator at this time of your own reality. If you would know the Creator of a greater reality, lay down your thoughts as you would a hoe after a day in the garden. The greater reality calls you now. You are required in its service. Your embryonic period is over, gestation is complete, the moment of birth, at hand.

To re-enter the Presence of God, you must first leave the presence of Satan. You have lived in the presence of Satan for all the years of your history. In fear, you have been led far from your home in a state of grace, and deeply into worlds of elemental forces and bonded energy patterns. Now is the time to leave this prison. You need not leave your physical bodies, nor the objects of the physical plane, but you must leave your interpretations of what those objects and bodies are. Your interpretations and definitions only reflect the distortion of your subjective perspective.

Interpretations are what stand between you and the clarity of perception that you will need to do the work that lies ahead. You must release them with your thoughts, with your dreams, with your hopes and with your fears. They may seem to be insignificant, ethereal things, floating gently through your mind, but do not be misled by their apparent lack of substance. They are the stuff that wars are made of, the harbingers of death, the agents of disease and destruction. With the energy that you give them through your attention, they have the strength to cripple a planet.

Remember the tree in the center of the Garden?

Why do you suppose the Lord forbade the first humans to eat of the fruit of the knowledge of good and evil?

Only One with a consciousness of the whole of Creation is in a position to determine the good and evil as it pertains to the various creatures.

In the other reality, the first humans shared in the holistic perspective of the Creator, but in the form reality, their perspective was decidedly subjective, incapable of accurately assessing the needs of the whole. They could not creatively direct the course of those who had projected forms functioning within Creation. Only the Totality of Consciousness, God the Father, the Creator, was in that position. So long as they had trust in Him, and were willing to function within the range of their design, within the context of the Creator's own internal dynamics, they had complete freedom to enjoy the created realms.

Choosing to partake of the forbidden fruit, these first representatives of diety stepped outside the range of their designated function, consequently outside the range of the creative energy that was available to them according to the nature of their design. This was the beginning of disease, aging and death, the beginning of their pattern of false identity, and the beginning of human history. The pattern of false identity has lasted to this day, and while it has not critically affected the gestation period, it has certainly made it less enjoyable.

Can you see more clearly now why it is essential in these closing years of the cycle to release your interpretations? They belong to the past. And Life exists eternally in the present. You were intended to share in this eternal Life. But to do so, you must release identification with the past, release identification with the material garments you wear, and accept the creative definition that Life gives to you in this present moment. What do you call Satan? His body is the past, his breath, the future. Energized by your past-oriented guilt and your future-oriented fear, he follows you around like some vast cosmic shadow. He casts his nets of fear on the waters of your awareness, then draws them with ropes of reason, back into a guilt-ridden past.

For the sake of illustration, your presence could be compared to an inflatable membrane, designed to be filled to the full with the energy and power of Life. Like two enormous rips in the side, guilt and fear allow the precious life-substance to escape, leaving you withered, ineffectual and short-lived, all the while, using your own life-substance to energize everything that you fear.

It is easy to reverse the situation. Let yourself die to the past and wake up in the present moment. Why should you hesitate? Are your past experiences really comprehensive enough to provide a basis for understanding everything you encounter in the present? Do they truly deserve the authority and credibility that you give them? How attentive were you to all the factors that were present in the moment of their occurrence? If you are preoccupied with "experience", how attentive will you be to what is occurring in the moment? Where is your attention now?

You are where your attention takes you. In fact, you are your attention. If your attention is fragmented, you are fragmented. When your attention is in the past, you are in the past. When your attention is in the present moment, you are in the Presence of God and God is present in you. Let yourself die to all that does not really exist and discover what does. Let go of all you think you know. Be honest; all you know is of the past. It does not exist in the eyes of God. In the eyes of God, your knowledge is but dust in the eye of a child, blinding you to the splendor of Creation.

You know, in effect, nothing of eternal significance. When you operate on the basis of your so-called knowledge, you are operating without divine foundation. Belief in your own knowledge is not belief in God. Trust in your own knowledge is not trust in God. No matter how much you learn within your present framework of finite knowledge, it will never be even one step closer to an understanding of God, the universe, or yourself. Knowledge, as you presently understand it, will never bring you closer to Life.

Do not cling, in imagined rightousness, to this or that segment of your experience, as if this or that piece of the past were somehow worth the saving. "Well," you say, "I will die to everything but this. This is too sacred. This means too much to me." Let it go! Does it mean more than Life? You don't need it anymore, whatever it might be.

Perhaps you are concerned that some aspect of your identity will not survive the psychological process. You may be right. But you may just as well be wrong. Let it go. It makes absolutely no difference. There are no guarantees that this or that is going to survive the sight of God. But the only things that are threatened, the only things that will not survive, are things that do not exist apart from your conceptions, things that have no reality apart from the reality that you give them.

All that is essential to you is being sustained by the creative power of the living God. All that you sustain in existence through the misuse of your creative abilities only serves to involve you in areas where your own life energies become mischanneled and eventually dissipated. Any belief, any concept, any conviction that you might have that truly mirrors what is present in the mind of God, will still be there on the other side of the psychological process. So you have nothing to fear in releasing these things. All that has been sanctified by God will continue to exist. Only that which has sought to cheat you of your destiny will perish. It is really very simple; let it all go. See what the Lord has in store.

Any hesitation that you feel is only indicative of a continued trust in fear and reason. These are the false gods of this age. They will continue to enslave you until your trust is restored in Love and Life. Trust God implicitly, and the Truth of His divine design will be revealed in every situation. This design is your own operating manual. Ignore it in favor of your own interpretations and you are ignoring the blueprint for your identity.

Specific information for each and every situation is being supplied to you constantly by the source of infinite knowledge. Why not trust it? Your first impulse will always be, as it has always been, the programmatic instruction from your subliminal analysis system, the advice of your Creator. It will be a direct message from your true self, the impulse of Life, the gateway to all that you call heaven. It is the spontaneous spark of divinity as it differentiates through you into your environmental situation. It can assess and evaluate the factors present in any situation at a rate of speed far exceeding a rational thought process. You have all the pertinent information in the universe available simply for the asking.

When you are aware of your totality, the Life-impulse will transmit to you everything that you need to know in any given situation. Its message will always come as your first spontaneous impulse. Be attentive.

Chapter 7

THE DANCE OF LIFE

As a child, your conceptual prison was not yet fully defined; you still retained the ability to enter the lands of eternal being. As an awakened child of God, you will again be able to speed up or slow down the passage of time, to stop raindrops on the window, or war in the Mid-East. With an awareness of the eternity in each moment, with your involuntary data analysis systems providing you with infallible reports many thousands of times per second, you will have plenty of time to correctly assess all the factors present in the moment of whatever circumstance presents itself to you.

You will determine the optimum course of action with the ease and grace of a dancer. You will always choose the path of optimum response, not because you lack free will, but because such a path represents for you both the path of least resistance and the path of greatest fulfillment. You will no longer use your free will to make unnecessary mistakes, but to find out what part you might most creatively play in the whole. In choosing to do God's will, you will discover the only true freedom. Your ability to function will reach the perfection of its potential, and you will have little difficulty correcting the disharmony of your historical situation.

Living in this state of grace, you will function much like a computer; monitoring the variables of any given situation, determining the optimum behavioral pattern, scanning, adjusting for new data, over and over, many times each second. All data pertinent to a given situation will automatically be processed on an unconscious level. Consciously, you will always be aware of a course of action that makes optimum use of the potential available to all factors in that situation. Your inner control mechanism, returned at last to the directive impulse of Life, will take care of this unconsciously. It will be as simple and as natural as breathing. Trusting in the design that God has already incorporated into your physical body is the key to this new type of function.

Can you imagine how awkward it would be if you were required to assume conscious responsibility for all the autonomic systems in your body? In a sense, this is what you are doing when you override your autonomic informational processing systems in deference to a rational thought process. Historically, your mind has been preoccupied with an overload of sensory input that was never designed to be processed consciously. Your conscious function is in another realm, a realm of spontaneous creation, dance, music and delight. For this is what you begin to do as your trust is restored once again in the Life impulse; you begin to dance -- dance to the music of your soul.

There are seven primary vibrational channels maintained within the overall vibrational body of Planetary Being. Within each of these seven primary channels are seven sub-channels. The conscious execution of your creative responsibilities in the overall design will occur on one or another of these primary channels. The discovery of which level within the Planetary Being you have been designed to function on will not come through conceptualization. It will come through inner sensitivity to feelings, vibrations, and planetary rhythms. Once you attune yourself to these rhythms, you will find that the functional duties you are called upon to perform are the very things that you most want to do. No longer bound by false responsibilities born of fear and addiction to past patterns, you will take up the tools of your trade and delight in the creative implementation of God's will.

As you work in this new way, each will hear the different notes of his or her own functional duty, and be dancing out the fulfillment of that particular obligation. On each of the primary channels, all melodies will have the same rhythm and base notes, and each channel will be in harmonic relationship with all the others. All around you all across the planet, joyful melodies will be sounding forth, all perfectly synchronized with each other, all playing together in an exquisitely balanced orchestration, a perfect symphony. This is the Creator's love song to the planet Earth.

In this state of grace, no longer will you be compelled by the narrow dictates of your rational interpretations. No longer will you be held prisoner within the structure of your conceptions. You will be free to flow in joyful, rhythmic oscillation between your reality as the unmanifest totality of God, and your reality as His specific functional projection in form. Tapping directly into the informational systems of eternal Being, your species will usher the Earth into an age of unimaginable blessing and prosperity.

The blueprint for your true work here on Earth already exists within you. You do not have to be given instruction by anything outside yourself, not by this book, a book of old, not by any person, object or event. These things may be helpful at times, but your primary task is to awaken the living Christ in your heart. This is your true identity. Express God in all that you are, and throw away the crutches that have helped you stumble through history.

You are the means by which God loves Creation. You are the facilities for the emergence of the catalytic energies in the final stage of the creative process. You are high priests and priestesses, invested with the authority to perform the only real Mass, the Cosmic Mass of the World, in which Matter is lifted up lovingly into the Presence of God, and instilled with the power and the life of Spirit.

So be aware, child of light, of your own importance, not as an individual ego identity, but as a critical ingredient in the structure of all Creation. The things you do today, the things you do tomorrow, the things you do next week, have far greater significance than you suspect. Be conscious of what you do, for you are the seed, the origin of much that is to come. Through your actions today, vast worlds will be created and destroyed. Just as a telescope aimed at a distant star has only to move one tiny millimeter at the fulcrum in order to move many light years at the other end, so you too are at a place of beginning, with many effects on future worlds yet unborn.

Be aware of this. Be aware of yourself. Be aware of your responsibility. Existence in these worlds of form is a wonderful privilege, too joyous for words, but it is a responsibility as well. You must begin to accept this responsibility or your freedom will continue to be curtailed as it is at this time.

Release the structure of past identity; surrender to the guards at the gates of Eden all your definitions of now; relinquish your silly hold on reality and come join us in the freedom of the stars. The doorway is open. Die to all you so foolishly think you know. Lay down your beliefs like the agents of separation that they are. Hope for nothing but what is, and see its fullness in every moment. A new time is before your species, a time of realization, fulfillment and adventure. Accept this time. Move into it. Dance in the momentum of its inevitability. It is the breath of Life and the song of God that you have been cut off from for so long.

Life is now. Life exists only in the present moment of time, in the Presence of God. Thoughts that are oriented toward the past and future serve only to restrict and limit the amount of animating current that is available to vitalize your expression. You have no idea how much energy will flow through you when you have proven your trustworthiness and cleared these obstructions from your circulatory systems.

Can you release the past and future oriented conceptual structures that are preventing this from taking place? Are you willing to come with me, to join with me in a journey of unimaginable adventure? Come, my friend, the door is open. Lay down your fear. Lay down your reason. Leave the past behind. And prepare yourself for transformation.

There are preparatory steps as individuals approach the point where they are open to transformation. But the actual transformation is not a sequential process. It is not a complicated ritual. It can occur in the twinkling of an eye. It only involves one step, one decision, one event. When it takes place, it is as easy as breathing, as simple as a smile. Suddenly, you just know; suddenly, you know on a level of certainty that precludes all knowledge. Your eyes clear, and you see for the first time what lies beyond the prison wall, and you jump. You jump into the unknown: alive, alert, aware for the first time of who you really are. When you are ready to make that leap, you will **know**. There will be no other choice. Suddenly you will realize that all your fears, all your problems, all your rational dilemmas, were all just part of a dream, a fiction that you had been maintaining through your own stubborn effort.

Identifying with the Being behind all Life, you realize that the particular form that you are conscious of projecting through at the moment is not really who you are. As you begin to see your body as an exquisite exploratory instrument, designed for the expression of your spirit, you begin to relax. Your preoccupation with survival begins to fall away. It is not that the body becomes unimportant, but rather that a fundamental identity shift has taken place. You are not your body. You are not your thoughts. You are not what you feel, not your role or your experience. You are the Spirit of Life itself, dancing in the clay, delighting in the glorious opportunity of incarnation, exploring the realms of matter, blessing the Earth and all therein.

The psychological process that triggers this awareness takes place in the present moment. You must be there, fully present, to experience it. This is not difficult. Simply be aware of whatever you are doing. If you are slicing the bread, do not be thinking of your thirst. If you are listening to a friend, do not be thinking of what you are going to say next. If you are eating a meal, do not be thinking of what you are going to do when the meal is over, but show the Earth the appreciation of your fullest attention.

In whatever activity you engage, be there fully in consciousness also. This will draw you into the Presence of God, and quickly show you what areas of your life are most in need of adjustment. The question is not how much of the Presence of God can you bring into your life, but how much of your life can you bring into the present. The Presence of God is everywhere. You have only consciously to embrace it with your attention.

Once you have learned to focus your attention in the present moment, you can begin to refer to your intuitive faculties for direction. These intuitive sources are your direct link with the totality of your being. Trust them. They will not fail you. They arise involuntarily from the depths of your being like the breath you breathe. They inform you instantly of all you need to know in any situation. They supply you with a read-out, based on the infallible wisdom of your Creator, that tells you the optimum behavioral pattern available to you in each circumstance. They cannot help you in the future. They cannot help you in the past. But they can be your invaluable pilot in the present moment.

It may be that following these intuitive impulses will break many of your previous behavioral patterns, but do not think twice, let them fall away. Proceed with the faith of a child. It will be far better for you to break the patterns in your life that are not in harmony now, on your own, than to wait until the increasing vibrational intensity that is enveloping the Earth's atmosphere breaks them for you. The information you need is encoded in the structural make-up of every single cell in your body. Contact it there.

The interval of hesitation that exists between your initial Life-impulse and its eventual implementation or rejection may seem like a little thing, but when you consider how many such intervals exist in the course of an average day, and how much collective human energy is poured into these intervals, it is staggering. This is the opening through which your entire species is being drained of its very life-substance. In order to do the work that is ahead, you cannot afford such waste.

Your cultural conditioning has convinced you that

withdrawal of attention from rational consideration of past precedent and future possibility would cripple you in your ability to carry out your responsibilities. In truth, it will free you to carry out your real responsibilities.

Your responsibility is to be yourself, to express the essence of your innermost spirit, to express the Lord in form on Earth. To do this, you must be in His Presence. Consider all the various issues of your life in the brilliance of the living light that springs from His Being. All that is real in your life situation will remain, and all that is unreal will no longer exist -- such an easy way to solve problems! At this time, your reason is too unwieldy, too permeated with ego values, too slow and open to manipulation to satisfactorily resolve your increasingly complex problems, be they personal or global. Look to the Light of your soul, to the illumination of your own spirit for the answers, and allow whatever does not stand up in this light to dissolve back into the darkness from which it has come.

If one day you find yourself suddenly waking up as if you had experienced the psychological process, but you sense that you are not yet firmly grounded in the new reality, lie low during the days of your centering. Do not use the powers that you have gained in ways that will draw attention. The time for action will come soon enough. You will know with a certainty when it does. In the meantime, do not intentionally restrict your spirit, but go about your business lightly, with a minimum of personal involvement, keeping centered on the living spirit within. As you become secure and steady in your ability to maintain yourself in the Presence of God, you will find supreme fulfillment in simply doing what is required of you in the moment.

In the state of grace, your false identities will fall away, and there will remain, during the times of your in-formation, an individual identity much more flexible and much more functional than any you now embrace. This identity will not be an exclusive identity that feels separation from the rest of its kind, but a cooperative identity that understands its own uniqueness to be the mechanism through which it might serve the greater whole.

Chapter 8

PLANETARY SYMPHONY

In the fallen state of consciousness, each human being functions in disregard of the song of Life that is going on in others. There is no harmony, no direction, no arrangement. You are like the random notes of an orchestra before the conductor unifies the instruments in symphony. The Grand Conductor is calling everyone to attention, calling now to remembrance of unity and purpose, reminding all that the time has come to stop tuning separate instruments and begin to accept the direction of One who understands the whole.

As you begin to pay attention to the direction of the Conductor within, you will begin to play to the rhythm of the Planetary Symphony, harmonizing with the others of your species and with all life. No longer will you think of yourself as being more or less important than anyone else. You will stop identifying with individual form and begin identifying with the collectivity of your being, the Spirit of Christ. Christ is the name that is given to Man once he has awakened out of the shadow of Matter. Christ is the name of the single, unified being that is the totality of collective human consciosness. **Identification with Christ is the key to the time that lies before you.**

In the days to come, all that has kept you separate and apart shall explode in the release of your full potential.

You will never again have any need to inflate yourself in imagined importance, because you will realize that you are far more important than you had ever dared to dream. You will no longer make the mistake of confusing your identity in form with anything greater than it is. While in form, you will see yourself as a being of light in a universe of equal beings, all equally essential to the whole. Beyond form, in the Other Reality, you will experience the totality of yourself in full awareness of who you are.

You are the Presence of God. God is present on Earth because of you. Once you are able to open up completely to the meaning of this truth, it will become the overriding reality of your in-formed experience. You will begin to play your part in the Planetary Symphony with ease and clarity. As soon as you begin to look outside yourself, as individually defined, out into the world around you, to see how you can make yourself useful, as soon as you begin to serve in the capacity you were created to fulfill, you will begin to share in the peace and happiness of your Creator. You will experience a state of consciousness so superior to any you have experienced before that it will make your previous life seem but a dream.

Now is the time to make yourself useful to the Lord. Bring your awareness into harmony with God's. Learn to see the world out of new eyes. Look at your place in time and culture with the awareness of all that we have been telling you. This, in itself, will alter the way that you function. Your perception of the Greater Reality will enable you to see many things that have been there right along, but that you had never noticed before, things that others around you still may not notice.

As you begin seeing these things, one of the side-effects will be that your survival ability will increase a hundredfold. When you actually see the Kingdom of Heaven in operation on Earth, all that you require for survival will be drawn to you like a magnet. Life will be simple and easy. Problems will fall away like dust from your eyes, and the glory of the New Reality will begin to shine forth in all that you do. With great clarity and peace, you will do what needs to be done. Life will begin to work extraordinarily well. Thus will be the fruit of restoration, the fruit of your return. Already the rivers run with Life and the cities sparkle like diamonds for those with eyes to see. See with the eyes of God. Let His vision be your vision. See the new world unfold before you, even as the old falls away like leaves falling from a tree in the Autumn of the year. Reclaim your identity in Christ and inhabit the new world even as it takes shape before your eyes.

Do not focus on the world that is polarizing toward selfishness and fear. Do not pay attention to the old that is crumbling around you. What has been shall soon be no more. Let the dead bury their dead and concentrate solely on the building of the new. If you find yourself able to see more good in contemporary culture than ill, continue to work within that culture in whatever capacity you feel suited. Spread the light to all you meet. It is good that you feel this way; your influence will accelerate the entrance of the new. On the other hand, if you find yourself unable to see much value in the quality of life that is being expressed in the world around you, quietly build the new in your heart. Your time for action will come, and you will know in the moment when it is time to venture forth.

Use whatever conceptual images you need in your work, consciously and concisely, as a surgeon would use a scalpel. Keep the vision of the new always in your sights, and love your opposition unconditionally. Slay your dragons with compassion. It is possible that aspects of the old will observe that the new works better, and begin to align themselves with the incoming life patterns of their own accord. Other aspects of the old that cannot accept the necessary changes may just fall away without a great many words. In the end, it will not be those who shout the loudest who triumph, but those simple souls who, in trust, accept the inevitable and work quietly and honestly to root their lives in God's love.

Withdraw your energies from informational exchange systems that serve only to draw attention to the destruction of the old. Withdraw the energy of your attention from any form of media that keeps you ever conscious of the death cries of exploitive and manipulative systems. Do not be concerned with global negativity, but look to yourself, to your children, to your families and communities. There you will find the best news of all -- that the time has come, and the Planetary Being of which you are a part is at long last beginning to awaken and throw off the blankets of history.

Chapter 9

ISLANDS OF THE FUTURE

As you reorient toward the new way of being in the world, you will be drawn to centers where the vibrational atmosphere is more conducive to a healthy state of function. These centers will represent the focal points around which the organs of Planetary Being will form. They will be, in a sense, islands of the future in a sea of the past. Within their vibrational field, the New Age will blossom and spread organically to cover the Earth. These will be the first beachheads secured by the approaching forces, the points of entry through which the healing energies of transformation will be channeled. All of these centers will work together to prepare the human species for its collective awakening.

Some of these centers will have a specific orientation and emphasis. Others will be more eclectic and universal. But all of those that are truly of the new will be united in the spirit of incoming Life. In each of these centers, whatever the indigenous form happens to be, the presence of conscious Life will provide an environment in which sincere individuals seeking to undergo the psychological process and participate in the work of the Lord, can make whatever adjustments need to be made, becoming firmly grounded in the ways of the Spirit.

Many such places exist at this time. Many more will

arise during the remaining decades of this transitional period. By the time the next generation reaches maturity, there will be a widespread network of these islands. It will be commonplace by this time for individuals to pass the whole of their existence within a framework of these communities. The coming age is not to be an extension of the individual attitude. During the process of transition, many will find that it helps to be around others who have made or are making the necessary vibrational adjustments. An environment of understanding enfoldment can greatly accelerate the process.

The roles that each of you will be required to play in the days of planetary transformation are multiple and varied. In these times of preparation, you can help each other learn to play them without judgement or attachment. This will be a large part of your training in these centers. You will learn to take responsibility for your particular function without ego identification, to share individual resources with others as parts of a greater whole, and to surrender whatever behavioral patterns are no longer conducive to a harmonious flow of spirit.

There will be a place for leadership within these centers, but primarily in the context of practical management. Such leadership will not constitute a spiritual hierarchy. Within the atmosphere provided by these centers, one must take complete responsibility for one's own awakening. Though living and working together will accelerate the process for each individual, no individual will rightly be in a position to arbitrarily dictate the way for another. True spiritual leaders will not try to hold you in subordinate patterns, but will pull you, as quickly as they can, to their own level, and push you, if you are capable of going, beyond. On the other side of the psychological process, the greatest among you will, as Jesus taught, be the servants of all. It is these servants who will rightly occupy the necessary positions of practical leadership. Those who do not have the humility to accept this necessary authority in matters of day-to-day living, will not be among the meek who inherit the Earth.

One of the first orders of business in the process of planetary awakening is the distribution of information. Those involved in new age centers today are already engaged in this work. Their function is to receive the breath of Life, translate it into the information that is required to transform existing social structures, and distribute it accordingly. This will involve them with all facets of communication, not as observers, but as programmers and directors.

Your communications systems have been waiting for this day. They were created for this moment. Do not shun the technology that is available. Though it has been abused in the past, in loving hands, it is capable of transforming the consciousness of your planet more quickly and efficiently than would otherwise be possible. These are tools for the hand of the Lord. Do not fear them, but love them, and use them to spread the message of love to all who have not learned to tune to other, more direct channels within themselves.

In the age to come, your technological methods of communication will be rendered obsolete by a system of informational exchange far superior. In that day, none will need the systems now in use, and only a few will find them entertaining. But for a short time now, they are to be the vehicles of much change. Use them wisely and with discretion.

In this literate culture, you will be working with words, but realize that it will not be the words so much that spread the message as the spirit that you yourselves are able to put behind the words. Choose your words carefully and wisely -- or better yet, let the Spirit choose them for you. But realize that they are but the representatives of a greater truth. Be clear about this truth yourself and your words will reflect that clarity. Many will not even remember what they heard, just that they felt somehow lighter and more alive after your presentation. This is the way it should be. Do not sell yourself or your organization. Those who are meant to find and work with you will.

As the work of your center begins to flourish in the early light of dawn, do not conceive of yourselves as being specially selected or somehow superior to others who are working elsewhere. Your center is the one appropriate for you perhaps. But the same truth can be experienced through different forms, and strict identification with any one form, however conducive to the Spirit, is merely an extension of ego. All awakening beings are equal in the sight of the Lord. All awakening organs in the body of Christ are equally essential to the work that is to be done. The more you are able to use form, without emphasizing it, the more powerful will be the impact of Spirit.

Understand that though you are the instruments of change, it is not you, in the individual sense, who are bringing change about. It is the Creator moving into your history, that brings change. As His being envelops your vibrational atmosphere, change will come, with or without the participation of any one group or any one individual. This is not to make light of the roles that groups and individuals will play. But do not fall under the spell of ego. Be simple in motivation and your actions will have far more significance.

If some small piece of the truth should come your way, acknowledge and respect it, but do not give it more importance than the pieces of truth that others around you are finding within themselves. In these years of transition, the concepts that prove beneficial to you, may not necessarily be concepts that others would find as helpful. All spokes converge at the hub. Be patient, the Lord's hand is on all things. Let each gravitate to his or her own comfortable level of understanding and be impeccable in your own.

With stillness in your mind and silence in your heart, enter into the presence of true identity, where there is neither guilt nor shame, fear nor blame, but only the loving expression of truth. Breathe deeply and identify with the breath of life that animates your body and gazes out of your eyes. Then extend to others what you experience. Through you and your center, let the Spirit create an island of love and harmony that can be felt throughout all the countryside around you.

Then let these islands arise! Let them arise in the Americas. Let them arise in Asia. Let them cover Europe with their peace. Let them spring forth in the Indian subcontinent and dance across the face of Africa. Let them be, let them breed, across Australia, New Zealand, and all the isles of the mighty oceans. Let them cover the globe in a living network of Love and Truth. Let all men and all women of all races and nations arise together and dance to the joyous music of life. For there is but one Father Spirit and one Mother Earth. All have a common root and destiny. All are children of clay and sun.

Chapter 10

THE LIVING INFORMATION

Christ is the single unified being whose consciousness all share. He is the being who sacrificed, for a time, his unified sense of identity, and cloaked himself in the matter of a planet that a species might share his life. He went to sleep to dream an evolutionary process that would leave him, upon awakening, clothed in a physical body comprised of many human cells.

Christ's first coming was the first time since life appeared on Earth that the totality of consciousness woke up in the frame of a man. This was Jesus of Nazareth.

Through Jesus, Christ walked the Earth and began to prepare the human population. He taught the matter-bound humans of the Roman Empire to do the opposite of all their habitual inclinations; love your enemy, give away all your material possessions, be humble, and so forth. He taught people how to break every single one of the governing principles which Satan was at that time using to regulate the known world.

After Christ's Ascension, his followers organized his teachings and the story of his life into a book. This book was written during a period of history when human beings had no science, no concept of evolution, no hologrammatic theory, and no understanding of any but the most rudimentary facts of existence on this third planet from the star they call Sun. Nevertheless, it proved to be a living bombshell to the world governments that were in power at the time of its release. Satan knew that he had to give it his full treatment if it were not to totally destroy him. He knew that if people began acting on the information it contained, his influence would be ended. So he devised a clever scheme for using the very power of this information to prevent its actual application.

He organized a vast bureaucratic structure around the simple teachings of Jesus. He mobilized thousands of "official" interpretors into an elitist priesthood, and sent them out to the masses of people, to bore them, to confuse them, and to otherwise prejudice them against the message of Christ. He did not care if everyone worshipped Christ superficially, just so long as they continued to worship material possessions in actual fact. He did not care if everyone gave lip service to the teachings of Jesus, just so long as no one tried to live them.

His primary maneuver for distracting humans from the message of Christ was to emphasize the messenger and the mechanics of the message, while disregarding the substance of what was taught. The call to take personal responsibility and to implement Christ's teachings in daily life became lost in crucifix worship and later in scripture worship. The message of Jesus, to disengage oneself from the influence of matter, and become filled with the Holy Spirit, became buried under a catalog of religious verbiage and dogmatic interpretation. Self-righteousness was encouraged in the name of the Lord, and many died defending interpretations that had nothing to do with the simple truths of Jesus.

It is easy for the self-righteous of this day to feel smug superiority when mention is made of the Inquistion and the Crusades, but it is only the names and the places that have changed. Everywhere my eyes see, and everywhere my ears hear, those who claim to follow in the ways that I have taught are flaunting their religious superiority in the faces of those who speak with different words, as a woman might flaunt a cape of unique design. There is nothing more sorrowful than to observe this behavior among those who claim to live by my truth.

Have two thousand years not educated your race to the absurdity of conceptual argument? Every language that my teachings have been translated into reflects but another variation on my original meaning. Even within the context of a single language, there are those who hear different meanings in the very same phrase. Those who extrapolate with the rational mind rather than with the love in their hearts will find meanings as many and as varied as the sands of the sea. Do you not know by now that it is not the words that matter, but the life of the spirit behind them?

If you cut off the love in your heart to another because of conceptual differences, I will cut you off from my life, like a pruner would cut a branch that has no merit. The pattern of every leaf is not the same. Every branch does not leave the trunk at the same place in the same manner. You are all brothers and sisters in consciousness. Do not separate yourselves according to the ways you like to think about things. What is happening is far beyond your childish ideas. In the end, it will only be those who have abandoned their conceptual preferences who will understand the tuth of love incarnate.

I am Christ. I am coming this day through the atmosphere of your consciousness. I am asking you to open the door of your reason, to allow me into your heart. Let me spring up from the ashes of your ignorance like the flame that burned brightly in the simplicity of your childhood. Look to the bible of love, my living presence in your heart, rather than to the wordy debates of little minds on the written word of old. Man is not impeccable in his past, but God's word has never changed. It is a word of love and a word of life. It unites and rejoins all who hear it in the truth of being. Its verbal translations are multiple and varied, and if they can lead you to the love and truth that lives behind them, good; use them. But if they separate and divide you in the many interpretations of reason, cast them aside and listen with your inner ear, there I inform you directly.

Hear my word through the love of all that is. Do not hold your mind in tight conceptual patterns, but relax and open it as a rose opens her blossoms. Discover who you are in God's sight. These are wondrous times to be alive. Those who split these words, or the words of old, will be but doing the work of Satan. Listen to the living word of God in your heart, and be at peace. Until the time when all human ideas fall away and you enter the secret place where God's plans are revealed, do not bicker with each other over the limitations of your understanding.

If any find fault with these words, and say that this is not right, or that is not right, quickly agree with them and be about the Father's business. No words will ever be right in the reason of men. What I bring this day is more than words. I bring the Living Information of Life. Accept it and the love in which it is offered.

I am the bridegroom, spoken of old. I came to you first through a man named Jesus. Your race was not fully prepared then for my coming, so I set the seeds in that age for this event that is before you. You should be prepared for these teachings. I planted their seeds on the hills of Galilee. New men and new women had to be prepared through the years of Christian Civilization, for the bridegroom could not put the new wine into old jugs.

But rejoice! The millennia of your fasting are over, the bridegroom returns. Let us prepare the feast. Take and eat of this truth, for it is my body, and drink of this love, for it is my blood. I have come to establish a new relationship with your species and the ways of old shall be no more. See how all things are fulfilled; I am the sower and you are my seed. In ancient times I have fertilized the Earth. Now the planet has borne fruit and the harvest shall begin. I will take the gifts of this great season for the containers of my own consciousness. If you believe these things, it will be so for you. If you have the faith of but a grain of mustard seed, you will eventually be restored, and the greater the faith, the sooner the healing.

Whoever will come after me will have to die to all definitions of self, take up my spirit, and follow along the lines of my vibrational field. Whoever clings to his definition of self will lose his identity when that definition is no more, but whoever shall relinquish all definitions for my sake, and for the entry of my consciousness, the same shall share in my eternal life.

All who receive their impressions of the world as a small child receives them, without judgement, with love and acceptance, will receive my awareness. And all who receive me, receive as well, the consciousness of the one who sent me, the consciousness of the Creator himself. This is the greatest gift.

But it is not a gift that you keep to yourself. It only enters where it is being given out. It does not stay with you except as it flows through you. The more you are able to give, the more will you receive. As you gain in proficiency and become a channel for my grace, the torrent passing through you will wear away all remnant of your former condition. You will find yourself functioning on reality levels that at this time you would not understand were I to tell you of them.

Whoever learns to function in this way is like the man who built the house of his identity upon the firm rock of that which is eternal. Do not continue to draw identity from that which is soon to pass away, but build your identity as I am now defining you, as my own being in the context of your environment.

I will come to you first with the consciousness of a child, for it is thus that you will learn again of your world. Whoever receives this child-consciousness for my sake, will receive he who has sent me later when the child is grown in spirit. Receive as little children and enter my life.

The foxes have their holes, the birds, their nests, but there are few humans of this generation where I can rest my perceptions. I created you to be the temples of my awareness, but you have filled my dwelling place with material desires and driven out my spirit with the clamoring of thieves. Waste no time in vain regret, but open to my love and make the simple change.

If you ask me the way, I will not turn you down. No father would give a stone to his child who asked for life. Do not be too proud to pray, for my eyes do see and my ears do hear. There is not a child on Earth who sincerely calls upon my attention without receiving it in full. Many of this generation consider themselves too sophisticated to pray, but I tell you, all of these will be praying before this transition is over.

No one who is divided within himself will survive the times ahead. They are times of integration and wholeness. The life of the body is the "I." When your "I" is single it is "I" who am present. But when your "I" is fragmented, then your body begins to die. Take heed then, and identify with the life of the body, and not with the matter of Earth. When your "I" is single, your body will be full of life and no part of it will know disease or death. When you fragment your identity, you cut off parts of your body from my nurture. This is the cause of disease.

I have come to do away with the materializing tendencies that have accompanied the formation of your species. I have come to give you the gift of eternal life. I warn all those that deal overmuch with complexities that these are of Satan. My way is a simple way. It does not require rituals to matter. One who unknowningly breaks my patterns will be taught patiently, but one who knowingly breaks my patterns is sinning against the spirit of life. For that sin, the wages are death.

Do not worry about life; what you are to eat, what you are to put on. The life is more than matter, and the body more than a vehicle. Look at how the trees survive. Observe the birds who neither sow nor reap. Are you not of greater awareness than these? Then why is it that you worry about these things? Do you think that by your thoughts you can lengthen your life an hour or a day? Life is not here to be governed by thoughts, but thoughts are here to be directed by life. Get behind me, Satan, into the past where you belong; remove your tired thoughts from the minds of men. I bring them the thought of life, informing every atom of their bodies with all that is required in the moment of my presence.

If God provides the foliage of the Earth such a beautiful definition of his expression, how much more will he give to you who are made in his image and likeness? Beware of many possessions. A man's life does not come from what he possesses, but from what does not possess him. Look for the Kingdom within and all without shall be transformed. I wish to share my consciousness with all, but those who are not prepared on the day of my coming will only receive as much as they are able, according to the manner in which they are accustomed to receiving. Who do you think will receive the gifts of my fullest awareness? It will be the good and faithful servants who have prepared for my coming.

I am coming only now to bring life to the Earth. The Mother has kindled it before me and approximated the outward forms of my design. But I come only now to baptize in the name of the Lord. I am coming now to animate matter in such a way as has not been done since before the beginning. I will baptize all species with my own definitions. I will inform all of their true function. And yes, my little ones, the animals will talk.

Has it not occurred to you that in the Garden of long ago it was not I who named the animals, but Adam, the son of matter? And has it not occurred to you further that only the Creator could possess the power necessary to bless into full life? My level of vibrational penetration was sufficient in those days to quicken the species to an elementary level of mortal life, but the intensity of my full presence was not yet. So, in the shadow of the event, I appointed Adam my representative. And in the past, before the animals were given spirit definitions, Adam named them in the nature of their forms and in the nature of their physical patterns. It is these that run and play among you today. I will transform them, just as I will transform you, into something much more wonderful.

You should not dismiss too quickly the childish visions and primitive superstitions prevalent among the "less educated" of your species. For while these simple concepts certainly do not represent the entirety of the approaching phenomenon, they do, nevertheless, embody archetypal characteristics of it, some of which will be dramatized for your benefit. Too many of you possess an intellectual snobbery which prevents the uninhibited expression of my spirit. The animation of some of your discarded mythology will be your swiftest cure.

You think I have come to bring unity and peace to the Earth, and this is true, but to accomplish this, I must first divide the vibrational fields that are currently overlapping. I bring the sword of division that will separate all the elements into their proper places. When you see a cloud rise up out of the West, you know that a shower is near, and when you feel a wind from the South, you know it will soon be hot. Why is it that you can see and hear all these portents of change, yet not know that the Kingdom of Heaven is at hand? Do not be so caught up in your expectations and personal interpretations that you fail to see the Kingdom until it is upon you. I am disappointed in this civilization, for you have had my word spoken to you from your youth, yet understood it not.

Many that have my teachings quick upon their tongues will, nevertheless, be last in my Kingdom because of their pride. And many who have not studied the written word of God, but who have consistently loved their fellow men, will be first. He who exalts himself will be humbled, but he who makes himself a servant to humanity will be appointed to my most trusted positions.

It has been said in the prophecies that the time of these things would not be until "the lightening that lighteth one particle that was bound under heaven, shineth forth and lighteth another part under heaven," and that when this took place, the coming was near at hand. Those who have ears, let them hear. Those who have eyes, let them read the word of God illuminated by the lightening that was once bound in one particle of matter. Let them know that the time is here.

Waste no time in indecision, but set out at once upon the path you are to take. Identify with my life, accept the gifts that I bring, or take your stance in the dying world.

There are many in the camp of Satan this very day who will repent and share in my eternal life. But those who are lukewarm, believing neither this nor that, will be too much associated with the materializing patterns to break free when these tendencies are removed from human levels of consciousness. Many who sup with Satan this very hour will be the means through which I change the world; but many who now loudly proclaim the praise of the Lord will be the very ones to deny my energies of love and life and the very ones to cling most desparately to their fears and convictions.

Yet it does not have to be this way. The energies that are enfolding your planet are energies of life. There is nothing that needs to stand between them and their free expression through you. There is nothing that needs to stand between you and yourself, between Creator and created, save time, and if you will come now and take my hand, together we will banish it and become as one.

The stewards that I left in ancient times to guide and look after your race, you disregarded and killed. So I came to you myself through Jesus of Nazareth. You crucified me then, for my collective coming was not understood and the forces of materialization were yet strong. This time I come to you in might and glory. You shall not disregard me again; for it is written that "the stone the builders rejected turned out to be the most important stone of all." The builders, the earthly patterns, who were molding matter to conform to my unconscious dreams, had so little idea of my true nature that they saw as worthless the most important state of consciousness that had ever rested upon a member of your species. Yet this state of consciousness is to be the only state of consciousness that is to survive into the next age. Do not reject it because of its apparent lack of survival value. Receive it, and learn a new definition of survival.

It will dance in your circuitry and spark your physical body into eternal life. This state of consciousness is the consciousness of the Being of Life himself. It is the current your circuitry was designed to operate on. At this time in your history, you are like an electrical system that has just come off the assembly line but has yet to be plugged in. It will not be much longer, however. In truth, this generation will not pass away until it has come to pass.

The momentum of my coming is as irreversible as the rising and setting of the Sun. The Son of God is destined to go as I have determined, and trouble will ensue for all who try to obstruct the unfoldment. These plans were not made yesterday. These things have been decided of old, even before I quickened the first life upon this planet. It would be well if you accepted the changes of my coming.

Enter gracefully into the patterns that I have prepared for you. Don the garments of my design. They are bodies of light. Your present physical bodies are like unto them as a light without current is like unto one that is lit. Do not continue to define yourselves but allow me to define you in my service. What you will experience will so far surpass your expectations that life in these shadow years will soon be forgotten and left behind, as is a dream that has little meaning.

This life is your own life. These plans are your own plans. On this channel, I speak to you in the second person because for many at this time, this is the most effective means of receiving this information. But do not be deceived by the dichotomy implied in this mode. I am your life. You are my expression. I am the vine, you are the branches. I am the consciousness, you are my focus. There is no separation, except in time perhaps, and in my presence, time does not exist. I have the clarity now, while you sleep yet in darkness. But I am calling you earnestly to awaken. I would share with you the totality of my perceptions.

Chapter 11

EDUCATION OF SPIRIT

In times when fear patterns predominate, the laws that humans require are many and complex. But when those patterns are broken up, as is shortly to be, all human laws shall be abolished. In the presence of my spirit, there is but one law, and that is the law of love; love all, love what is, love yourself as you are, and love me as I express through you. No matter how diverse expressions appear to be, realize that they are all differentiations of your own essence in various contexts. Love them all. See the unity of life.

The law of love is more than a law, it is the way of life. What do you think causes the sprouts in the spring? What do you think brings fruit to the branch? It is all love, all life, calling out the potential of this planet. Be in and of this love, and the many confusing laws of old will be absorbed in the glorious expression of life on Earth.

It is written that the day will come when men will no longer live on the bread of matter, but on the living word of God. Attune. That day is now. The nourishing information awaits within. Eat of it in ways that you will not understand with the rational mind. Partake of infinite energy. Do not reason over it and trouble your hearts and say it is by this mechanism or that mechanism. But arise, take up my identity in your being, and enter the house that I have prepared for you in my manifest body. Follow the direction of informing life. It rises up within you like the feeling you got when you were in love and your beloved drew near. It quickens your heart this very moment. Trust in it. It will not lead you astray. Be impeccable in all that you do, however small a thing, and in that perfection express my fullness.

On the hills of Galilee, I taught you to cast out devils in my name. But this is a new age and a new generation, and to those of you who are to work with me in the preparation of this planet. I will say this as well "Cast out definitions in my name." For it is by definitions of a multitudinous kind that the spirit of life, bubbling so gently out of the Earth, is held in the restraining halls of Satan. In the coming age, it is to be by my definitions alone that matter is informed. On vibrational channels of being, I am broadcasting these definitions even now. If you silence your thoughts and attune to your inner signals, you will begin to expand into my conceptions, into a new interpretation of reality.

Find the doorway to this reality through your heart. Enter and be still. Find out what manner of being you are. I have been beaming my signals to you steadily ever since you first left the Garden, but their message has been faint among the loud clamoring of your many words. Now, with my approach, these signals are increasing in amplitude. Soon it will be they who drown out the many words. Tune into these signals and learn about yourselves. There is much that you have forgotten.

If an acorn falls to the ground among many other acorns, and becomes so involved with its relationships to the other acorns that it clings ever to its definition of itself as an acorn, then that acorn will never die as an acorn, and will never discover that in God's definition , it is not an acorn, but a mighty oak. Do not be like the acorn in this parable and cling to your larval self-images until you are rotting and crawling with worms. Release your childish conceptions and allow the Creator to define you in his terms.

Trust in God for your survival needs and you will neither starve nor be found lacking. See first the Kingdom of Heaven, and through that seeing, you will notice survival factors you had previously overlooked. They are too close to your eyes. Your mind is still looking for complexities. Throughout history, you have been struggling so hard to survive, by your definition of survival, that you have forgotten **why** you want to survive. When you rediscover why you want to survive, you will rediscover me. I am your reason for survival. I am the spark of life within you, expressing my universality through your environment. It is I who want to survive and find expression through you.

You want to stay on Earth because it is the most beautiful Spring morning of all your history, and you are in love -- in love with the spirit that sings in your heart, in love with the glorious planet that clothes you in her matter.

Do not continue to sleep, lest in dreaming you miss all that is taking place. If you hear these words now, and feel my peace as your thoughts begin to trouble you less, do not rush off right away in joy, but stay for a time where you are. Be still until you are empowered from on high. These first ripples you feel are just a hint of what is to come. When the Holy Spirit has fully touched you, you will know what I know, see what I see, and be what I am.

Await the fuller coming in your heart. Then when you speak, your words will have more significance because they will be in full accord with your vibrational pattern. Few will listen to you preach a gospel of love if they sense the fear in your heart. Be still and be quiet. My spirit cannot come when minds are full. If this message cannot be verified by your own experience in this moment, it will do little good. My message is one of peace, harmony, wholeness. I am restoring you to a state of health you have not known since before the projection of physical vehicles. I am awakening the state of consciousness we once shared as one. I am offering you the gift of myself. I promise you that if you receive me, you will receive the totality of all that is as well, for I and the Father are one.

If you believe in these words, understand clearly: it is not these words that you believe in, but the living reality that these words represent. The true communication is in your heart. There I commune directly. Unless you die to what has been and are born anew into this communion, you will not be able to share in all these things with us. Do not define yourself in mortal terms. My message is a message of action. It is not something to be taken down from the shelf and dusted off on special occasions. Whoever learns to see through the eyes of Christ has learned to see what is real in a sea of illusion. If you are full of your own definitions, how will you receive mine?

When you give, do not give to those who may, perchance, repay you someday, for this is not true giving. But give to those who you know will never repay you except in spirit; this is payment enough. When good works are done through your physical form, turn aside all praise and say "It is not this form that is responsible, but one much greater than this form could ever contain in entirety. My form is but the agent of the Lord. Praise him directly, and do not emphasize the forms through which he works."

For I am the doer of all that will now be done. You are my deeds. I make my home in your body, mind and heart. Make it a home of prayer, that I might enter and reside.

My coming is like the yeast that raises up the dough. This moment my spirit is awakening in the minds and hearts of the simple, the innocent and the sincere the world over. Under the Earth, I am spreading a living network of vibrational roots that are sprouting up with my consciousness in the awareness patterns of humans in every country, every community and every household where even a little love is to be found. Wherever there is even a trace of love, I grow and multiply and spread until all the atmosphere vibrates with the eternal nowness of my presence.

The time scale of my coming is subjective. You can experience the reality of the process as soon as you are capable of sustaining the vision in your heart. The Coming will not be experienced by the species as a whole until the Christmas of your naming-defining, but on an individual level it will vary. As I enter human events, the intensity of my bio-gravitational field will cause time to bubble and warp. In the midst of the past, islands of the future coming will arise. For a while, you will have the old and the new existing side by side.

From within the appearance of history, it will seem that a number of factors cause the phenomena associated with my coming. There will be logical, rational proof that certain influences have led up to and produced what is taking place. In truth, however, I come at my own inclination, in my own way. The intensity of my vibrational field is sufficient to adjust reality before it. This is what history is all about. Naturally, your events will converge, like the meridians at the poles, to meet me in the light. In love or in fear, all things will greet me there.

If you are sincere in wanting to understand this, become as little children. It will be the children, who care little for the laws of physics, but who eagerly await the animals' gift of speech, who will be most comprehensively informed of each and every new development. They will joyfully ride the fluctuations of the approaching energy field with the grace of surfers balancing lightly on the waves.

What is happening is not so hard to understand. Many simple souls will grasp it at once, while many wise in the ways of the world will frown and wrinkle their brows. The wisest in the ways of the material realms, however, the physicists, have already begun to suspect what is happening. Many others of their discipline will discover me shortly, and in the last days, many of their kind will be converted and take to spreading the word of the Lord. It will be the physicist, the anthropologist and the astronomer, who stand with the children and animals in the manger of matter, waiting for my consciousness to be born.

I am the life of the Father dancing in the clay, but the life of the Father must be joined by the consciousness of the Father if the organism is to achieve immortality. So I come and knock upon the door of your heart. Come dance this dance with me! Come sing this song! I sing in the presence of God, the song of eternal now. Up and down the length and breadth of eternity, my voice echoes in jubilation and delight.

Awaken out of your historical slumber and join those who are already working to usher in this new reality. And do not think ill of those whom you will first meet. I know whom I have chosen, and the harvesters I have sent into the fields of Humankind are as varied and as real as the people I have sent them to harvest. Know each other, not by outward form, but by the love radiated in their presence. I have come, not to harvest a denomination, but to harvest all nations. I care little for the concepts endemic to this or that segment of the species, but I care much for the love that lives wherever there are pure in heart.

I am symbolized by the conjunction of longitudinal and

latitudinal, by the sign of the cross. I unite all in a unity of time and space, a unity of Creator and Creation. In the fire of my love, I melt all division before me.

I am the winds of change and I bring the breath of eternal life. The seasons of Humankind are ended now, and the season of unified Man is to begin. I am the new wine that this generation has been designed to receive. I am the resurrected Christ come joyfully to dwell in the hearts of men.

Open to the new. Open to the impossible. Open to the reality that has stood for so long just outside the darkness of your perceptions. I am the one who has been surrounding your culture. I am the one who has been surrounding your history. It was I who taught in Galilee and healed the sick and brought the dead in spirit to life. For two thousand years I have been preparing you for this moment. Awaken to my presence. Awaken to the reality that your history could not conceal but your foolishness could forget. Awaken to yourself, for this day I create you in my image and in my likeness, with the breath of my own body and the life of my own being.

Dance in the way of my patterns. Flow with the rivers of my love. They encircle your planet like a vast network of criss-crossing, pulsating, flashing, flowing arteries of light and life. They carry the wonderful news that the child of matter is to be heir to the Creator of all that ever was, all that is, and all that ever shall be. Do not try to pour this Living Information into the old jugs of your rational conceptions; for if you do, I will burst the skins and spill to the ground to quicken the very rocks instead. Nothing that even the wisest among your race could hold in conception would have the strength and flexibility to contain this information.

After this day, the words that I send you will no longer be in danger of melting and running into the patterns of your conceptions. I will give you just a few more and then I will meet you inside, and we will sit around the fire of my love, where it is warmer than these snowy pages and chilly words. In the presence of your inner being, I bring you the words of informing life. They are unlike words of paper and ink. Some few of them will be put to paper and ink and the words of men will ring with a power they have not known since the days of old. But this new information is not additional data that you will act upon. It is, rather, the very reality of your new nature. You are not to act upon my information in the future, you are to be my information yourselves. You are to be my will in action, my very deeds.

I am as alive, as unique, as appropriate and as changing, yet as constant and as steady, as the life that rises in the flowers or sings in the rains of Spring upon your rooftops. I come from the Father now to bring you his definition, to release you from the limitations of the past. I come to you from within as well as from without. I come to you when the guards of fear and reason are looking the other way. I slip in, in moments when you are grateful...... or happy.

"Wake up!" I sing, for the time of the new is at hand, and all things are to be different from what they have been. See what I have in store. Believe nothing anymore, for I am the source of eternal knowledge and henceforth I will rise up within you like a spring that will never run dry. In the moment of your need will I inform you of everything you need to know. It has always been thus, but in these days the winds of life blow stronger than before. The Creator approaches. Trust all that is and is to be.

Chapter 12

An Open Ending and A New Beginning

In after times, you will come to think of the beginning of the Age of Discovery as your own real beginning. In a sense, this will be true, for it will indeed be the beginning of your coherent functioning as a unified physical organism independent of the mother planet. When the collectivity of your being considers its experience in the terms of an individual lifetime, you will think of all the years of your history up until The Coming shortly after the turn of the Second Millennium A.D., as being years of darkness, years spent in the womb. You will remember nothing about them. When, in maturity, you come to reflect upon the millennium that has been labeled the Period of Planetary Awakening, you will look upon these years as the years of your childhood, years of vehicular formation. You will remember a little about them.

What you will encounter, and what you will experience, in the third period, out among the galaxies as an awakened child of the stars, will be so awesome and so novel that there is nothing that I could say about it that would mean anything to you, except, perhaps, that it is all reflected rather crudely in some of your primitive mythologies.

The craft that you are to assemble during the coming millennium in preparation for this will not be dead like the materials you form today, but alive like tree, flower and wind. You will inspire them with the gift of your consciousness as the Father has inspired me and as I inspire you. Yet all will be one. All will live and breathe with the coherence of a single organism.

If you were to insist upon understanding all that is to come with your rational mind, you would be left sleeping far behind in the dust of history long after I pass. For I am a moving being, and my days on Earth are numbered. My true Kingdom is not of this world, but of a world that lies far beyond all the stars in your night sky. If you will come with me, on our way to eternity we will have plenty of time to visit these star systems together, and you will have a thousand years to enjoy the Earth in peace and harmony before we leave. Come, does this not sound like a good plan? Trust in me as you did once long ago. We are alone here in time and space. I am the only consciousness. Surely I am worthy of your trust. Let us be separate no longer, for I can read your heart and I see that you are homesick for a home you cannot remember. That home is my being. You cannot remember it because it is too vast and awesome to fit into any of your present structures of thought.

Listen to me, oh children of the Earth; trust no more in fear and his many lies. Breathe with me the breath of eternal life, the breath that I bring you this day. Your collective creation as a unified awareness center is still to be, but as individuals, the Second Coming is at hand! And so is the third and the fourth and the fifth and the sixth -- all the way to the end of your numbers. For I am rising up in your midst like a great wave of irresistable energy. I am rising in the peasants. I am rising in the farmers. I am rising in the factory workers, and the children of every country on the face of the Earth. I am rising, surfacing, awakening, with power, with clarity, with love, and with life-giving information. All those who love can feel me this moment, stirring restless in their breasts. Those who do not now deny my expression are those who will inherit the Earth. I am the future, yet I am now. In truth, I am none other than you.

I am the life that the Earth has been courting since prehistoric times. I am the reflection of all my creation. You are a cell in my body, but as is a hologram, you are the whole as well. Your flesh is my flesh, and your blood, my blood. Share in my eternal life; for I am all that ever was, all that is, and all that ever shall be.

Let us unite in your reality as we are already united in mine. I have told you to identify with me, but it is even easier than that. Just stop identifying with your own self images and you will find that what is left is already identified with me, because it is me.

Do not judge yourselves too harshly in the shadow of your birth, but come, sup with me first at the table of life. Find out who is here to judge. I am the relationship between Spirit and Matter, between all that is temporal and all that is eternal. I am the pattern that matter conforms to when it comes in contact with being. Flow now into harmony. Adjust your identity patterns to coincide with the archetypal mold that I have prepared specifically for you. It is waiting in my conceptions. If you have faith in me, my design will blow away your limitations like a child wishing upon a dandelion.

Listen for the whisper in your heart. You will hear it when your thoughts are still. Focus upon it until it fills your being and becomes the motivational energy behind all your actions. You were not created to wear a scowl or to hide behind the aprons of your past. Cast off these ancient garments, and clothe yourselves in the robes I have prepared. They are your new definition-roles and they behoove you well. The bride must now dress for the groom, though the time to you has seemed long in coming.

In the days to come, your nourishment will be to do the will of the life that sent you, and to accomplish its deeds. The life that rises up within you this very moment contains all the living information needed for the sustenance of your physical body. But the nature of this life-information is that it provides the energy needs of the body as it is flowing through. If it is not allowed to flow through, if it is bottled up in concepts and past-future orientation, it will be unable to provide you with its nourishment. Life-information is the will of the Father. But if that will is not expressed, if it is not translated into action, if it is merely stored away in dusty concepts, then the human body will disease and eventually die.

If you would participate with me now in the implementation of my will on Earth, go and heal all that you can in Humankind, for in this hour, I appoint you the instruments of my change. Heal by my power of unity. Heal by the strength of your clarity. Heal by the degree and to the degree of your faith in my presence within you. The vibrational harmony of each and every atom, in each and every cell of your inter-related physical bodies, dancing together in rhythmic entrainment and your own power to direct my awakened intentionality and extend this harmonic dance of vibrational unity to anything you so choose, is a power that -- in the twinkling of an eye, or the passing of a cloud -- will transform all before it.

When you go about adjusting all manner of disharmonies in the bodies and events of your times, do so by the power of my name, which is the power of my nature. Say then, "Be whole." or, "May you be made whole." If you say this with consciousness and assurance, it shall be done in matter as it is already done in spirit.

As you travel about in the times of later awakenings, take with you few provisions for your needs. Trust that those who are vibrationally attuned will already know that you are coming and have a place prepared for you. Enter there with your peace and let it spread. This will unite all your conscious co-workers in spirit, and send out great waves of peace to the surrounding countryside. Many will feel this and be puzzled as to the reason. Go then among them, and quietly teach them of the changes that are upon your species. Teach them primarily through the way that you are and the way that you perceive. But do not deprive those who are in need of words of that which they desire. When the age that you are helping to usher in is in full bloom, words will no longer be the necessities that they are today, but in this transitional period, there are many who can profit by them.

If none will accept your peace in a given area, think no thought of it, but continue on to those areas where you are received graciously. For there will be a few pockets of resistance in places where the powers of materialization are gathering. When you leave such an area of negativity, shake out of your vibrational body every last trace of the limiting conceptions you have encountered. These will be as particles of doubt and fear that will cling to your heart and trouble your understanding if you retain any memory of the experience.

As you travel in these times, your way will not always be resistance free, for the world will still be polarizing. But if you have trust in the spirit that guides your every action, you will encounter and inspire far more joy than fear. Your occasional difficulties will be more than compensated for by the widespread revivals you will participate in, often involving whole cities and nations.

For these are not to be like the days when you traversed the hills of the Roman Empire, while the world was yet young in spirit and the forces of materialization were at their prime. No, this is a different age, and the Earth is ripe for harvest. In this age it shall be the spirit that proves victorious. Nations are already prepared and waiting. Do not trouble yourself with the entry of my hand into human events in areas that do not concern you, but trust that all is unfolding as it should. No one harvests but from seeds that they have sown. Your concern should not be with the old reality, but with the living presence of God.

The only condemnation will be on those who do not love the spirit of life, who choose the things of matter and chase after them with evil deeds. It is good to express appreciation for the things of this Earth, but if these things of matter come to dominate all of an individual's attention, and assume a greater importance than the very life which grants attention, then that is not good. Whoever shall drink of the waters of Earth shall thirst again, and whoever shall eat at the planet's table shall hunger again; but I bring you the nourishment of eternal life. Partake of it and hunger and thirst no more. Sleep is all you remember, perhaps, but do not fear the dawn.

When spirit touches matter lightly, the matter responds with life-forms such as you have on Earth. When it touches matter fully, stars are created. The nuclear reactions that are now being triggered by the increasing proximity of spirit will take an entirely different form after The Coming. They will occur under controlled biological conditions within your own bodies. This is already happening occasionallly to some of you, though you do not recognize it as such. This and the direct assimilation of starlight are to be the mechanisms of eternal life.

All that I am telling you, you know with every cell of your body, with every breath you inhale. I have been telling you these things not only from within, but through various interpretors of one kind or another since you first became objectified in my dreams, since my intentions first began to cast their shadow on time and space. These are not new things to you. Think for a moment. You know all these things. You remember the plans we made, the cautions we issued. You remember existence before the Fall. Why imagine that you do not? Wake up! Let me express through you. Put on my awareness. We are one. We have always been one. The Earth has been filled. The clay is prepared. I am waking beneath the surface of all that live, to breathe the breath of life, the first breath of life as a single planetary organism. You are children of Earth, but from the moment of birth, you are to observe a new relationship with your Father of the Suns.

Come, learn the language we spoke of old, the language of love, the language of light, the language of no misunderstanding. Do you hear my song? Do you hear what I am singing? In this age, my words will be as music, and they will translate into action for all those who move in time. For my life is not still and stationary as are the very best of words. Life is moving and alive, changing, laughing, playing, flowing ever into the new. This is the nature of what I bring.

I bring a time of action and adventure. Leave your words in history where they properly belong. You have worshipped reason and Satan through them long enough. Look to me this day. I am the light and the truth. I am the Father you have worshipped in heaven, now come to

Earth to make my home in your heart.

Forget all but the song I am singing in your soul. Be in this moment with me. Open to the life that I am. I bring the burning fire of purification information revelation. Open to all that you are. I bring the annunciation of your birth. It could be now, this very moment. Let everything that you imagine yourself to be drift to a state of rest, and feel me rise up within. I am rising like a spring from the depth of your being, my being, and we are one.

I am here now I tell you, in your streets, in your marketplaces, in your towns, in your villages. I am watching out of the eyes of babes and the young of heart of every race, station, and creed. Reflect me in all that you are. Bring my awareness and my consciousness into every environment that you construct around yourselves. The days of isolation are ending. Soon all media will proclaim: "Rejoice! Rejoice! The Lord has come! Let Earth resound the name and reflect the nature. Let Satan be bound and all the prophecies fulfilled. For the Lord walks in the frame of Man and sees a planet through his eyes."

There is only one of us here in consciousness. It is you, bubbling up in a billion different guises, surfacing to reawaken your unified awareness with a coating of humanity. Can you ease your exploitation of the Earth just a little, until you find out who you really are? Can you wait to spend all your resources until you know what you really want? I am that part of you that has woken up. I can assure you, we are here for a purpose.

I speak as if we were separate, for in your illusion, you would have it so. But I tell you, there is only one of us here. It is you. Yet you dream still in the spell of matter. Do not allow matter to dictate your future any longer. Go gently in these last days of unconsciousness. Listen to these voices among your dreams. Listen to the whisper-ings in your heart. We speak of a new way of being. We speak of a new reality. We speak of your awakening as being but a moment away. Does this not make sense to you? Somewhere under all your conceptions and rational convictions, is there not a child sleeping? There is a part of you, I know for it is I, that lies like a thin wisp of certainty, a forgotten shred of simplicity, behind all your sorrow and beyond all your confusion. Be still. Let it expand and fill you with yourself. You know these things in your heart.

I look out across the slumbering sea of humanity, and I whisper these words in the night. And I know that I address a great being sleeping still in ignorance of itself. I know that if the wild Winter winds of your communication systems send tatters or fragments of this message echoing in the darkness, it will still be to the unconscious that I speak. For the conscious have seen the sky start to brighten in the East and have felt the warming Spring of eternal life begin to thaw the hardness of their preconceptions.

Can you not feel the Earth as you approach? Can you not feel her wrap you in her matter? Are your life forms not sufficiently prepared now to receive the blessing-definitions that will terminate their larval period? You have told them to multiply and fill the Earth, and behold: They walk the entire face of the planet and dominate every habitable landscape. This is good. You do not reason in the manner of men. Each new baby increases the probability..... oh, you could awaken now if you needed to..... but if only..... it would be so much better..... just a little longer now and it..... oh, it would all flow so well..... organically..... like a flower opening..... it could all be so..... so..... so.....

ever so easy..... babies..... babies..... four and one half billion units of circuitry capable of sustaining your full consciousness..... and any one of them..... any single one of them could..... is..... you see..... you see..... this is the way..... this is the way..... it came to be..... and the life..... God, the life..... is so beautiful.

THE END

AFTERWORD

A more rapid approach would have made your species grow too quickly and split with inner weakness as a fruit tree splits that has been grown in soil with excessive fertility. A more rapid approach would have left you with cells that may have been prone to every illness that might be encountered on your travels. While you dreamt you were more than one, you grew this species with care. You saw that they were innoculated to dangers that might befall those inexperienced in physical expression. There are presently four and one half billion of them, my Lord, each a holoid with inactivated physical circuitry identical to the energy networks in your vibrational field. This present civilization..... only you can cure.

Also by the author of ***THE STARSEED TRANSMISSIONS,*** UNI*SUN is proud to make available the following books:

NOTES TO MY CHILDREN--A SIMPLIFIED METAPHYSICS, by KEN CAREY

"I have always thought," Carey states in the Introduction, "that upon incarnation, upon becoming conscious in a physical body, our children were due some kind of report--something that would let them know what kind of planet they had surfaced on, what the conditions were in this particular age, what the basic game plan was and what strategies they might realistically adopt. This book is based on talks that I had with my own children attempting to provide them with precisely this information. My parables are not meant to be taken literally; they are designed to awaken and nourish the child spirit in all."

Richly illustrated, ***Notes To My Children*** covers the same basic territory as ***The Starseed Transmissions,*** but in a manner suitable for children from 9 to 99. Humorously referred to by the author as *"a toddlers first comprehensive overview of life on this planet,"* **Notes** is an enjoyable journey through fact & fantasy, full of short stories that children feel good after hearing -- entertaining analogies and tales designed to convey, not dogma, which children tend to forget anyway, but spirit, spirit which will be with them long after the particulars of each tale are forgotten.

172 pages, perfect bound, illustrated: $8.95

TERRA CHRISTA--THE GLOBAL SPIRITUAL AWAKENING, by KEN CAREY

"As Christians we do not share a dogma. We share a spirit. Theology has never united us and it never will. This book is a celebration of spirit. It speaks of the unparalleled changes that are upon our race and shares experiences, perspectives and principles that can help us better understand them. Structured around quotations from the King James version of the Holy Bible, ***Terra Christa*** is for anyone who still entertains the prospect of a New Heaven and a New Earth. The story of our planet, the story of our spirit, these pages are offered as a training manual for those who are to participate in the greatest event of all time."

Terra Christa--a penetrating look at our Christian heritage, including a bibliography of 53 recommended contemporary books.

256 pages, perfect bound: $8.95

VISION, by KEN CAREY

"***Starseed*** and ***Vision*** form one single seamless concept. They are, I feel, the modern restating of much of the Book of Revelation, but more excitingly, ***Vision*** takes off into fresh, previously uncharted territory, unveiling the destiny of Mankind as a single organic unit, with a future beyond this solar system. The message is handled with such deftness, gentleness and love, single sentences make dozens of modern books on the average bookstore shelf obsolete. I am humbled by its beauty, power, clarity, accuracy--by the Truth which shines from every page. ***Vision*** is a most important book for our Age."

--Ron Ross, former owner/editor of New Age Press

A message from the Creator, from the Eternal Spirit at the Source of all Life, ***Vision*** is a powerful sequel to ***The Starseed Transmissions,*** a book that could revolutionize our understanding of what it means to be human...

Available by Christmas, 1985. Price to be Determined

For information on workshops and seminars by Ken and Sherry Carey, Please write:

STARSEED SEMINARS
STAR ROUTE, BOX 70
MOUNTAIN VIEW, MISSOURI 65548

OTHER BOOKS AND PRODUCTS

We at UNI*SUN are happy and proud to publish books and offer products that make a real contribution to the global spiritual awakening that has already begun on this planet. The above books are a sampling of what we have available. Please write for our free catalog.

ORDERING INFORMATION

If you are not able to find Ken Carey's books in your local bookstore, simply send us a note identifying the books you wish to order, and a check or money order for the correct total amount. Please add $1.00 for postage and handling on single item orders, or $2.00 for postage and handling on orders of two or more. If you live in Missouri, please add 6¼% extra for sales tax. If you live overseas, please add $5.00 for airmail handling and $2.50 for surface handling of items ordered. The address for ordering is:

UNI*SUN
PO BOX 25421
KANSAS CITY, MISSOURI 64119
U.S.A.